HOUSE BORN OF MUD

HOUSE BORN OF MUD

A Builder's Story

William N. Gates

SUNSTONE
PRESS

SANTA FE

Sunstone books may be purchased for educational, business, or sales
promotional use. For information please write: Special Markets Department,
Sunstone Press, P.O. Box 2321, Santa Fe, New Mexico 87504-2321.

Book and Cover design • Vicki Ahl
Body typeface • Candara
Printed on acid free paper

———————————————————————————————————————

Library of Congress Cataloging-in-Publication Data

Gates, William N., 1930-
 House born of mud : a builder's story / by William N. Gates.
    p. cm.
 ISBN 978-0-86534-751-9 (softcover : alk. paper)
 1. Building, Adobe--Anecdotes. 2. House construction--Anecdotes. I. Title.
II. Title: Builder's story.
 TH4818.A3G38 2010
 690'.837092--dc22
                      2010008272

———————————————————————————————————————

Published in

WWW.SUNSTONEPRESS.COM
SUNSTONE PRESS / POST OFFICE BOX 2321 / SANTA FE, NM 87504-2321 /USA
(505) 988-4418 / ORDERS ONLY (800) 243-5644 / FAX (505) 988-1025

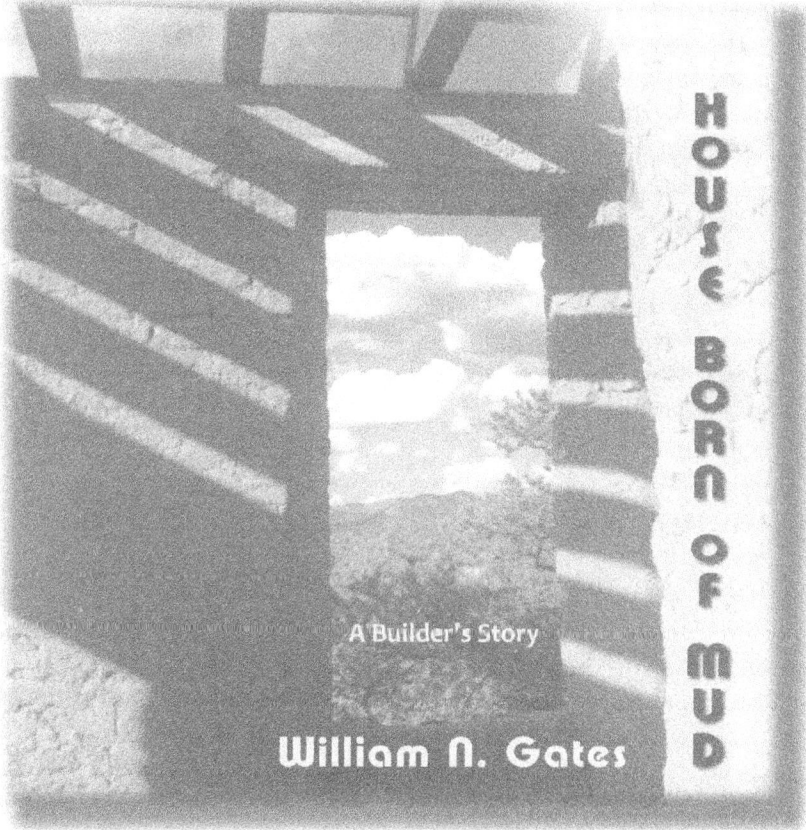

HOUSE BORN OF MUD

A Builder's Story

William N. Gates

Foreword

The house rises from its land like a butte or a mesa, presenting a long solid front with no windows and with one deep, shadowy entrance-cove. As was intended, it echoes the old Hacienda de Los Martinez near Taos, which was built in a quadrangle like a fort. Since the nineteen sixties it has passed through various hands; the present owners have lived in it longer than the original family, whose fragments are now scattered to far corners of the country. Different people have resonated to the place in different ways. This is the builder's story.

I

In October of 1962, my wife Ann and I bought a 17-acre piece of land on the rim of Arroyo Hondo, about seven miles southwest of Santa Fe, New Mexico.

We engaged Buford and Kitty Bartell to draw detailed plans and specifications around Ann's basic design of a quadrangular house with a patio in the middle. Although we would not begin building for another year, we chose a site on and over the rim of the grand arroyo. Our acreage dipped into the valley, with a view of mountains all around—the ridge of Cañada de Los Alamos to the east, the Sangre de Cristos to the north, the Jemez to the west, the Sandia to the south. The piñon and juniper that covered the site and steep banks of the arroyo stopped at the floor of the dry watercourse, and there began chamisa that covered the bottom like a river of pale green, sweeping and winding back through a dark cut toward dark foothills and mountains.

Our well was drilled in the fall of 1963. The dry mud from the drilling still flows down the slope; I hope it all washes down some day. Pat Mullins dug the well and built a crude, tiny well-house underground for the well-head and the 50-gallon pressure tank. Mullins was a short, stubbly man in a shiny helmet who spoke in a high-pitched voice that always seemed about to shoot out of control, always on the verge of a scream. Gerhauser the grading contractor called him "Squeaky." He dug Gerhauser's well and Gerhauser got fed up with him: "I run him off," he said.

Mullins' son Jackie did most of the work on our well, and he and his pretty wife and baby camped out on site, in a trailer next to the drilling rig. The place was littered when they left. Jackie struck water at about 250 feet, and got a flow of about 25 gallons a minute. With a piece of broken mirror he reflected sunlight down the casing and showed me the water shining at the bottom like a new dime.

Power was brought in that fall, and I put in my first hard labor digging a three-foot hole for the meter pole (later moved). I dug it right behind the small grove of trees off the northeast corner of the house. There was nothing on the land then, it was a cold day, and Michael, then 3, kept me company playing in the dirt and buzzing around. Lengthy dickering with the slow and immoveable Public Service Company finally resulted in the setting of only one extra power pole, at no cost to us. Once the electric hookup was installed, a line could be run underground to the well-pit, to run the pump and provide water for construction. Later, another line would be trenched to the house from the meter pole.

On that October day Michael gathered some firewood, bringing me twigs and proclaiming how good they'd be for fires. We piled it under the big tree that still stands by the garage. It is hard now to remember what the wild site looked like; almost with the first digging I began to forget. I think the ground was pretty bald except for a couple of junipers that used to be under the garage or southeast corner. There was one good clump at the site of the dressing room and laundry that had to be ripped out and several good trees under the dining room, and many under the living room. I regret all those trees and can still remember the size and shape of them and how they grew together.

The house stands half on what was the crown of the site,

and half over the edge and down the steep slope of the arroyo tributary to Hondo, a deep forested ravine which the children have named "Echo Valley."

The site was never staked out by a surveyor; Tom and I did it ourselves, and I am figuring out how we could have done it. We must have set stakes at the house corners and set offset stakes to pinpoint those corners after the true stakes were bulldozed out. Wall lengths we knew from the few drawings the Bartells had finished: 79 feet along the back, 70 feet along the side.

"Seventy-nine feet?" Tom said in soft disbelief. "Seventy feet? This is a big house. What do you want such a big house for?"

I rented what Tom called a level and tripod, a sighting instrument set on three legs. It had a scope with a crosshair, on a swivel that was leveled by screws on three sides, so that when correctly adjusted it turned full circle and remained level. Using the rod as a constant measure, you mark the rod at the crosshair elevation, and then you can sight on any number of different stakes, and with the rod find and mark the same ground-level on those stakes, and all the marks will be level with each other.

So, after the first elevation was marked on the rod, Tom sent me to another point—say the southeast corner—and turned the scope to find the rod as I held it, sighted the mark and had me transfer ground-level to the stake at that location.

The markings were made on the batterboard posts. These 2 X 4 posts were offset several feet from the actual house lines, so they would not get dug out, and they provided moorings for the batterboards and lines. Three posts bracket a corner, forming an angle, so that the true corner can be found plumb under the intersection of lines strung from them.

With one hand I held the rod to the post, and with the other hand clamping a yellow pencil horizontal across the mark, I slowly moved the rod up or down until Tom signaled that he had the pencil lined up with his crosshair. Then I held the rod still, took the pencil and marked the batterboard post right at the bottom of the rod. This was one of those times I wished I had four hands.

After the batterboard posts were marked at the same elevation, Tom took a length of 1 X 8 lumber and with my help leveled and nailed it flat across two of the posts, with the top or bottom edge of the board exactly at the elevation mark. To get the other side of the batterboard corner, we either sighted on another mark or used the board already nailed as a guide. He marked with a penciled arrow which edge, top or bottom, was to be used for the line. At the finish of this work, all corners were bounded by these angles of batterboard; and for a long time afterward they imposed the only precise definition on seas of rucked-up dirt. Not only at corners, but wherever we needed a fix on the outside line of the house, we had to put batterboards: midway down the line of the east wall, we set a batterboard facing west, to keep the line of south kitchen and dining wall; down the north side, another, facing south, to keep the line of the hall. I could squeeze my brains to give the splendor of the batterboards and lines. For a while they were the house: level fence-like angles and signposts that held the planes and crossroads of taut white cord. From those lines of cord depended the depth, straightness and squaring of all foundations and walls.

For groundbreaking, Tom drove a stake about midway down the east wall-line, so the bulldozer would know roughly where to start digging deeper; at that point the floor was to drop, and with it the crawlspace below it. Same at the dining room -kitchen, where floor-level was to drop another two feet.

Pine won't stand much pounding. We had to bash the stakes into very hard ground with Tom's sledgehammer, and the tops were all crushed in; some two-bys split clear down the middle. Tom asked me how old I was (he was getting tired of driving all the stakes). I told him I was 34. He was 45, with a wife and eight children. That was the first time he openly sought my help. The way he told you to do something was: "Why don't you bring some two-by-fours, if you want to." "You can start filling block if you want to." "If you want to" was the cushion, softening an order until it wasn't an order; yet you did it. He spoke in a soft voice, a diffident, downtrodden voice, with a singsong lift at the end.

The idea was catching. If I wanted a man to do something, I said, "Do you want to help me here" or "Do you want to give me that board." I usually took the cue from Tom and suggested rather than ordered. It's easier on the spirit. Things get done with better grace, better will.

None of the building trades is easy, and the workers live hard lives, working into their sixties and on to the day they drop. No breaks, half an hour for lunch. Time goes fast. Busy days go fast, Friday is upon you very quickly. Winters are "slow," employment is "slow," time is slow and miserable. At the end of such a winter, I first met Tom Weatherford, and he was low and in dire need of work.

I engaged him; but the beginning of the job was slack, uncoordinated, and Ann remembers him sitting by the hour in his car in front of the house we were renting, waiting for me, waiting for work. He wouldn't go in the house or take coffee or read the paper, only sit there staring grimly and wretchedly out the window.

As soon as work developed, he suggested and I bought a power saw with a 7½" diameter blade, a "Wizard" on $30 special sale at Western Auto. Tom always called it a "Skilsaw," from the main maker of these tools. He proceeded to make four sawhorses and a pile of stakes. To make a sawhorse, Tom took a two-by, notched the four corners on a bevel to seat the legs, cut legs of a 1-by, tapering them toward the bottom so they wouldn't splinter, and nailed the legs to the beveled corners of the two-by and braced them so that they were spread wide and stood firmly on the ground. "Burro," the

Spanish people called a sawhorse. After two years' hard service, his burros were worn, paint-spattered, and scarred like dog-chaws, and their legs began to wobble; so I had him make two new ones along with a cradle of X's for cutting firewood.

Such a trifle as a sawhorse required measuring and exact cutting and fitting, and it would have taken me hours just to figure it out. Tom knew what he was doing. Experience plus patient, steady application: his motions were never hurried or spasmodic, always cool and consequential. His pace was deceiving, because it seemed it would take him a long time to finish something; yet it never took him long. I would set him a task I thought would keep him busy for a while, and he would finish it in minutes. There were periods in '64 when no carpentry could be done, and I had him shoveling dirt like a day laborer, or laying adobes. He didn't like it but I paid him his $3.50 an hour just to keep him.

In the beginning he wore a cap and jacket, and until he got his new striped ticking overalls, he used lumber company aprons, short aprons with pockets and a hammer-loop. He wore ordinary crepe-soled low shoes and khaki shirts. Carried his tools in a long-handled wooden tote-box that he made. Tools are expensive; he used his until they broke, patched them up and used them more. Handles of hammer, pick, hatchet were very worn, taped and nailed where they'd cracked. Numbers, inches on his tape measure and his six-foot folding ruler were faded, and the joints were loose. What he didn't have he borrowed from his brother Carlos, also a carpenter. He always carried on him hammer, nails, pencil, ruler.

In the box: large metal square, combination-square (with attached 45-degree angle), blue chalk, chalk-box, chalk-line, several different screwdrivers, chisel, crosscut saw, hacksaw, keyhole saw, brace and bits, hatchet, small crowbar (octagonal steel bar with a wedge at one end for jimmying, a crook at the other end with a notch for clawing and prying), three-foot-long metal spirit level with four bubbles for testing both vertical and horizontal, plumb bob, lines of cord wound round pieces of wood, wrench, two rickety pliers, pencil stubs, whetstone valleyed hollow from use, 75-foot tape measure in a case, plane, rasp, nailset. All jammed into the narrow box whose bottom was full of chalk and sawdust. I had to pick my way carefully down through the massed points, teeth and blades to find a tool.

Tom smoked heavily and rolled his own, shaking tobacco along the paper, licking and closing the very wet, loose cigarette. One was always in his mouth, lit or unlit, like a scorched old pillow losing its stuffing. I still find everywhere the wrappers of cigarette papers, and Prince Albert tobacco cans, some so bleached that the red has turned pale orange, and the black-bearded figure faded to ghostly atoms.

The only things that pulled Tom up short were the idiosyncrasies of our wishes and the intricacies of the architects' plans. He never refused, only resisted. "Well, you don't need all that," he would say. "You want to go down thirty inches?" he said, as near incredulity as he would allow himself. "O no-o-o!" he used to say, laughing gently. By a searching melancholy stare, a mouth that stifled laughter, he would reveal that he thought we were mad. "You going to use that to make your ceiling?" "That's a pretty-looking board," he said as he turned over a specially grimy, weathered piece. In a man so soft-spoken, the least emphasis, twinkle or drawing-out of a word sufficed. He would crouch for a long time over a page of the plans with a dark horny finger on some outrage or other, and study and whisper and murmur in a mournful, baffled tone. He would whisper and mutter to himself while he worked: measurements, bits of Spanish, curses, "Chingar!" and "Eghh, " a sort of goat-sound of disgust. When he snapped the Spanish loud and fast, he was angry. But he rarely showed anger. Dirt would cave in on some wet, just-floated concrete and he would go "Eghh."

"O no-o-o! It won't freeze down there. There ain't no water around here!? Thirty inches? I don't know. Maybe it will. I don't know."

He would back up and deny his own objection, if it were an objection. So I had to find my way through his resistance, supple, pillowy, constant. Sometimes he won me over. Well—*vamonos*—

I still hear the little squeaks and chirps of the steel crawlers of the bulldozer in its rocking and tilting drive forward, blade shelving up another layer of earth. "Groundbreaking" is right. I remember the dirt peeling, lifting, swelling and crumbling like a wave of the sea, and the dust breaking off it like spray tearing away on the wind. In places the dirt turned up powdery, ashy, a light gray caliche dust, so light a gust of wind would tear great clouds and sheets off it. In other places the soil turned up clay, red and hard as wood.

The excitement began that April morning as soon as we spotted the huge trailer-truck coming toward us over the hills and dips of our drive. It carried a bulldozer on its back.

The driver backed the trailer up to a mound of dirt. Then the bulldozer operator drove the cat off the back until it dropped onto the mound. It looked as if it would tip over backwards, but the crawlers ran it off safely.

First it cleared trees. I felt a painful pleasure to see the evergreen tops sway and fall, and hear the roots crack out of the ground. The cat lifted them out as if they were little weeds. The operator never touched a tree without first looking to see if I wanted it out. He was a small man with an Indian look. It was nice to watch his hands at the controls, darting quick and sure from lever to lever.

All that day and one hour of the next he graded our site. Tom and I checked his depth by sighting through the level and tripod on a narrow marked board which T. called a "story-pole" or "story-rod." There were different marks that he put on it for different levels, whether brick or wood floor. For wood floor, we figured: the ¾" finish floor, plus ¾" subfloor, plus 7½" for the joist, plus the 18" crawlspace equaled 27 inches total for one mark on the rod. Where the house stepped down two six-inch steps, we added a foot for another mark at 39". Brick on sand floors had to be left higher in grade; in other words, leave dirt undisturbed as a base for brick. This was tricky, especially where the different depths ran side by side. The floor elevation of the kitchen was to be the same as in the dining room, both rooms three feet below that of the first two bedrooms; but the kitchen was to be brick and the dining room wood, with a cellar dug out under it. The bulldozing by necessity had to be a rough excavation. But even though I had to perfect the grading by hand, the machine did in a day what several men with pick and shovel might have accomplished in a week. Fourteen dollars an hour was the price.

Ann brought the children out, much excitement. At the sight of the mess, Sarah wailed, "They're ruining our land!" It looked a ruin, with mountains of torn-up dirt everywhere.

Michael kept shrilling to the cat-operator, "Hey, man! Man!" Finally the man gave him a little ride on the 'dozer.'

The Bartells came out. Buford wanted to know how I felt and how Ann felt, and I told him it was exciting as hell. Pleased, he grinned, one of the few moments during the whole course of this construction when he was pleasant to be with.

"*Seems* to know what he's doing," he said of Tom Weatherford and his work at that point.

The next stage was digging the foundation trenches. To prepare for the ditch-digger, the backhoe, Tom and I had to trace all the outside lines with lime.

The batterstakes around the future basement were miles in the air. In addition to the three-foot drop from the southeast to the northwest corner, the land fell away so fast we had to nail up extra lengths of stake, and because of wind we had to guy them with braces staked to the ground. Even so, I think we set the lines at their true elevation there, or they'd have been too high to work with.

(Let me backtrack and correct to avoid any more confusion. The batterboards were not in place at this stage, only the offset stakes, which were set in line with the true corner stakes to act as a guide after those were torn out in the grading.)

So, our next step in preparing for the backhoe was to reestablish and stake the true corners by running lines between the offset stakes. Once the corners were reestablished, there was no further use for the offsets. They were removed and the batterboards were set up, bracketing each corner. Using the batterboards, we adjusted the cord until it was plumb over the corners, and then marked or notched or nailed the edge of the batterboard where the cord crossed it. We checked squareness by first making fast two intersecting lines, then measuring off on line three feet from the intersection (the corner), then four feet on the other line, tying pieces of string at the three- and four-foot marks, and then measuring the diagonal between the marks. If the diagonal measured five feet, then the corner was square. It can also be done by measuring 6', 8', and 10' on the diagonal. Don't ask me (or Tom) why, just do it. It is a formula. Another method is to measure the diagonals between opposite house-corners; it requires a long tape. If the diagonals are equal, then the corners are square. Right? God help him whose diagonals do not measure up to 5, or 10. Fortunately ours did, so we didn't have to start all over again.

Once the batterboard lines were set true and square, we had ourselves a taut diagram of cord which was a plan of the house strung in midair, level and plumb over the building site. At any time this string plan could be projected on the ground with a plumb bob; and it was a removable plan which could be taken off and wound up around a stick, then secured again, just as tight and accurate as before. Wonderful thing!

This work fixed the two rectangles composing the L of the south and east wings of the house. There were also several insets, tucks and angles in the foundation, and also some interior foundations, such as the one wall dividing the brick and wood floors—a retaining wall to keep the dirt, sand and brick from the crawlspace under the wood. Also, a retaining wall between the future sewing room and the kitchen. And others. Those details all had to be traced by lines and batterboards, too.

With the lines up and drawn taut, we began projecting the plan on the dirt. Wind and dust blew; lines wouldn't hold straight, and the plumb bob swayed when we hung it from a line to drive a stake under the point. We would slide the plumb bob along the batterboard line and drive another stake. Between the two small stakes we tied a line, looping one end around stake one and tying it to stake two. This short line was kept close to the ground, but not touching, so that I could trace it closely on the dirt with lime shaken out of the pinched lip of a coffee can. Then, remove stake one and leapfrog it to the next plumb bob point, tie the line again and lime another 8 or 10 feet. Bob, stake and lime. I liked this work, it made such simple sense. By the time we finished, we had the outside lines of the house drawn in white lime upon the ground. We went home hoping the wind would not blow them away.

The Day of the Backhoe: early May, sunny, clear, and a hard cold wind blowing all day.

The family came out to see this bright yellow prodigy go to work. We were all enthralled. The backhoe had a loader on one end (a scoop like a bulldozer's), and shovel or "bucket" on the other; two motors, one to drive it, and the other to generate power to work the shovel. It looked like some giant jointed bug, on the order of a praying mantis.

In the hands of a good operator, it digs with great power and delicacy, no wider or deeper than you want it to. Without disturbing a crumb of dirt outside its designated line of work, it gives you as neat and square a trench as man or machine can dig. Case "Construction King" was the name of it. I imagine a folk ballad sung by Cisco Houston, called "Backhoe Bill," or "Nell She was a Backhoe."

The trenching started at the bottom end of the site, where the cellar was to be excavated.

Rrrmm—rmm—rm-rmmm: the motor of the digging arm had a hum of its own, a nice, rapid, deep even "rmmm" that was different from the driving engine. I can recall it clearly. It was not tied to gears like the other but went at one rpm, punctuated by little breaks, like throat-clearing. I heard it all day and it soaked well into my ears and memory.

The seat swiveled to face the digging controls. To set the machine firm, the operator, a very young, nice-looking man in a red baseball cap, lowered four compressor-operated legs with round flat feet, each of which could grip the ground at any level. Touching the bucket to the ground as a brace, he lifted the whole front end of the backhoe and swung it left or right to whatever position or angle he wanted. Then he placed the points of the two-foot-wide bucket inside our lime-line and began to dig.

Arm and bucket were a wonder to watch. The arm is hinged by heavy pins at three joints, and along the back of each segment, like tendons controlled by brains of machine and operator, run compressed air lines, one at the shoulder, one into the elbow, one into the wrist. The air lines pull or push those members. It was the most articulate machine I'd ever seen, so insect-like, so animate.

Slowly, carefully, with slight weavings and noddings, the arm reached out, way out, at least 20 feet at full stretch, and put its paw and its row of big claws on the dirt. The dirt was tough, couldn't be gathered right in, so the arm lifted and brought down the paw several times, tapping hard, and the claws punched and mauled the dirt until it broke. I can hear those claws scrape the dirt, and the ponderous, hollow "dump, dump" of the bucket as it punched ground and loosened rocks. After a pile of earth was loose, the paw took it away. It was a backhand paw, it didn't push, it pulled; came down on dirt contracting, curling in on itself and cupping out a load that would probably fill two wheelbarrows. When the arm lifted, it drew back its forearm to keep the paw from spilling its load, then swung sideways, reached out and uncurled the paw to allow the dirt to shower down.

It took the backhoe half the day to dig the cellar, nine feet below dining-room-floor elevation, plus four inches for a slab floor, plus ten inches for colossal footings. I checked the excavation with a story pole while Tom went on marking and liming at the other end of the site, at the angle of the future hall where the construction was to be temporarily closed.

When there was dirt to be cleared, the operator raised and shut off the bucket arm, raised the four bracer-feet, swiveled his seat around and used the backhoe like a tractor. He lowered the scoop and turned it upside down to push and plow with; he turned it right-side up to fill it with large amounts of dirt and to move same from place to place.

Then he started on the trenches, with the backhoe straddling the line and moving backward piece by piece. It is best at trenching; all the two-foot-wide ditching was done that afternoon. He came back next morning for an hour with a 16-inch bucket, to dig the smaller footing trenches. Nine hours in all, $10 an hour, cheap at the price. I and two other men would have killed ourselves for days and weeks shoveling it all out. When I think of just moving the dirt, pitching it shovel by shovel out of the cellar hole ...

The fun was over. Now followed the perfecting of trenches,

steel-reinforcing and pouring of concrete footings, all hard, grinding toil. The two of us did it all.

The footings were to step down to follow the drop in terrain. The deepest drops came along the outside wall, from kitchen to basement. Trenches and steps had to be trimmed and cleaned out. I worked hard for a day or more just on the cellar and footing steps leading down to it. The outside wall footings were planned to be continuous and connected; therefore the big dive into the cellar to join the cellar footings.

(Before I get further: story poles for checking the backhoe had to be figured and marked according to pumice-block units, and multiples of the 8-inch block, as follows: 10" for the footing, plus 8" for the block height, or 8" X 2 (16"), or 8" X 3 (24"), or however many courses or "rows" of block needed to raise the foundation wall above ground level. In addition, Kitty Bartell insisted that to guard against freezing, footing-bottom must be at least 30 inches below "finish grade" or ground level. So we tried, but did not always succeed, to observe and incorporate that stricture into our figuring, too. Also, at wood floors, Tom had to figure in not only block height but wall-plate. (l½", a 2 X 4 laid flat), joist, subfloor and finish floor. The foundation height, and digging depth, had to be calculated to catch the wood floor at its designated elevation.)

The risers of bearing steps (dirt steps of trenches) had to be cut according to block height: 8", 16", or a maximum of 24"; and the treads must be at least two feet long. Foundations must be laid on unbroken earth, not on fill. We had hard clay earth in most places, so walls and floors of trenches held shape without forming. But in the cellar hole, about two feet below the surface we struck a deep stratum of sand and gravel that went all the way to the bottom; so cellar footings had to be formed. Sand kept showering down on me, on my hat, down my neck, as I worked. The wind overhead raised whirls of grit and dust down there. It was hard. Curses and groans. Try to cut an even wall and it would piddle away. Step too near an edge and it would cave in and have to be cleaned out. It grew impossible to shovel the dirt out of the cellar, so I had to run a plank down and wheel it out. A strong fir plank, under the weight of me and a full wheelbarrow, it bounced nervously and bent like a bow but never cracked.

I balked at the size and extent of the footings planned by the Bartells. Tom said we'd go too deep and use too much block, and—we didn't need to? I stared at him dumbly. He disclaimed, "Maybe you do, I don't know."

The plans were printed in blue on white paper. Rolls and rolls of paper, 11 sheets in all, though we only had a few to start. Even the first few drawings, foundations and wall-sections were complex and baffling. Drawings came in plan and section, with swarms of numbers, symbols and terms in shorthand, and arrows stabbing in and out. Later I could read them easily, but then I had a struggle. Being so new at it, I wanted to do everything right. Tom told me I followed the plans too closely. It was a problem of whose word I was going to take. The drawings were a maze, and it was a trick to learn which to consult for a certain detail—floor plan, foundation plan, wall-sectional drawings, house-sectionals, house elevations, not to mention the specifications, the "specs." Each shed a different light from a different angle on the same problem.

Aside from the loose sheets supplied to us as we went along, we used up two complete sets of drawings, and the third is now so dog-eared I'll have to get another. I had to learn and use all the jargon of architects and builders—"lay up," "rebar," "aggregate," "bearing," "set up," "darby," "joint," "grade," "finish," "rough-in," "casing," "sash," "waste," "valve," "hydrant," "hose bib," "bulkhead," "backfill," "buck," "jamb," "header," "on center." Not to mention suppliers' euphemisms such as "water closet" and "commode."

Plumbing drawings were not supplied until long after the foundations were laid and walls were going up. The Bartells farmed this work out to an Albuquerque engineer named Earl Ryan, and he was a stone in my shoe. They gave him the job in February or early March and he frittered over it. It was not until late June that I finally had plans and specs in hand, so that I could apply to plumbers for bids and get the work going. I'm not sure now that Ryan did right by us. On his butt in an office 60 miles off, never here once, no understanding of site or wind, only adding and calculating room-sizes, walls, et cetera, and feeding it into his BTU computer! "Thermal Units!" They must be set out in some engineer's rulebook, how many BTUs are required to heat so many cubic feet of space. Bah! Now we have a house that won't get warm when it's cold outside and a wind blows.

Contractors such as Ryan tend to pass the buck. You say the heating system he designed is not working well, he says it's your fault for not building a tight enough house, or it's the plumber's fault for not installing the system right. You go to the plumber, but he swears he did everything he was supposed to, so it must be the designer's or the supplier's fault; and never in all your searching will you find a single person willing to take responsibility.

The Bartells should have kept after Ryan smartly to produce the plans. But they didn't. They wouldn't. Weeks and months bled away while I was wringing and twisting for lack of the plumbing plans. I remember getting this message from Buford: "Remind Bill that there are lots of headaches in building a house."

Buford had a stickling, professional purism. He got touchy if you didn't state precisely what you meant in trade terminology. His speech was like the language of the specs, which he wrote. (Kitty did all the other work, though she signed his name to the plans.) It was as if one could conquer the chaos and uncertainty of expression through exact usage and correct trade cant. Architect's specifications have a rigid, uniform wording and phrasing, like Army manuals.

Buford seemed a man under great pressure, under fire. He was 44 years old, tall, thin as bone, emaciated-looking. Lank, stringy, graying hair fell over his forehead. His nose looked worn down, low on the face. He had long, hollowed cheeks and a thin mouth; his eyes were shielded by shiny steel-rimmed glasses. It sometimes seemed the stresses of his life were pulling him apart.

Kitty kept a grip on her emotions. Intensely cool, yet somehow gracious. The same fires that have scorched him seem to have left her more cool, sane, graceful. The calm not of a corpse or slug, the calm imposed by a fierce will, was in her voice and motion. Spanish-black hair, glasses, white teeth, pretty when she smiled. Lonely, wanting to talk, hard to get her off the phone; on and on in one long sentence, voice low, even, almost monotonous; suppressed violence, of humor, or wrath, in her voice.

They were from Texas, and had the accent, not coarse but faint, moderated: "brayss" for brass, "pahn" for pine. He said "hayf" for half, "ayrea" for area, "Sayrah" for Sarah. "Get you a piece of two-ba," she said. "Get you a mop."

Since both were architects, they had to suffer from the home-job conflict: she with her will to practice her profession, he with his will to keep her at home, to be mother and housewife. I always tried to deal with her, not him. She was here, he was away, working at a new job in Albuquerque. She knew the plans, she was pleasant. Yet she deferred to him, he was "The Architect," the final authority; his name was on the plans. But I called him only when I had to. I dreaded his criticism of the work done by others, which I must answer.

He was stingy with praise, generous with dispraise. Neither he nor Kitty would say if the idea of the house seemed good to them. He only said he wanted to design the house that we wanted.

(Maybe that's a good rule as well as a good dodge.)

I recall him inspecting the plumbing progress. Sewer lines were plugged, all piping filled with water, to put pressure on the system. "Looks like a leak every damn joint," Buford said, feeling, peering. If he chewed out Willie Sena, I was not there to hear; but he might have, because Sena looked pretty sick when I drove up. It seemed they'd had a showdown.

He crawled under the floor, took hold of a run of copper pipe and shook it. It rattled loosely. Sena had not secured piping with hangers. Bartell found sewer pipe blocked up off the dirt on hunks of wood. "Hell of a job," said Buford. "Half-assed job. Carpenter ants get at the wood, pipe falls down, you've got a break in the joint."

Sena wasn't there, but I was burning, stung by my first suspicion that I'd better keep a close eye on Sena, that Sena would make a monkey of me whenever he could.

Looking back over my pay records it seems impossible that so much ground-work and footings were done in four weeks. But Tom and I did it. I know Tom must have figured and worked like hell. If I'd known more, I could have been more help, but I too strove like hell just doing what he told me.

Footings were to be reinforced with steel rods running through the concrete; block foundation walls were to be locked to the concrete footings by vertical steel rods set in the footings and rising up through the block. Twenty-foot lengths of steel rebar, delivered by truck, had to be cut and placed along trenches, bent around corners, bent down steps. Called "number 3" or "number 4" according to diameter (3/8" or 4/8"), the round gray reinforcing bar was ribbed in rings all along its length, to help bind it in cement.

The rods gave a dull, steely, ringing clank together. They gnawed the sawhorses when laid across them. Hacksaw raised a shrieking song as it cut steel, and glittering dust dropped to the ground. Rods were harsh on hands, blisters formed quickly. Some of the verticals had to be bent like an L to give them a hook into the concrete. One after another, we calculated length according to block dimensions, marked the steel and sawed it halfway through. Then I stuck the rebar into a length of pipe, stepped on the pipe to hold it down, and, grasping the rod in both hands, bent it upward with a quick yank. Usually the rod would snap clean off at the saw-cut, and save you sawing it all the way through. I used the same pipe to bend the ends of verticals for the cellar footings. Other verticals I cut straight, to be stuck in the floor of trenches at intervals, to be pulled up into the concrete later, after pouring.

For all this reinforcing work, we followed the footing detail drawings. Horizontal rods had to be laid in the trench so that concrete would surround them; and had to raised from the trench floor, set in from the sides and spaced apart. Bricks were good blocking—a row laid flat at intervals, especially at corners and steps where it was hard to keep rods up off dirt. Bricks raised the rod just about the right height. Without brick or stone, rods could be suspended by wires from a board bridging the trench. Rods were tied end to end with wire to form continuous runs of steel. Short spacer bars were tied across them at intervals, to keep them apart.

Bulkheads to stop or step the pouring had to be formed. Tom made these forms out of boards held together with cleats nailed across the back. He had to set them plumb and square athwart the trench, and stake them firmly to the banks, 10 inches above the floor (footing depth). These forms were the risers of steps so that when you ripped them out later, you would have steps of concrete behind them. Forms would act to block the flow while it filled in the space behind them.

Extra-tall bent verticals for the cellar had to be tied to the horizontal rods and braced up from the lumber footing-forms. Here there could be no trenches because of sandy soil, and Tom had to form the footing trenches with 2 X 10 planks. With extensions, those vertical rods must have been eight feet tall.

Cellar footings were extravagant, 28" wide, and 42" wide at the areaway. Kitty explained that it was necessary because the walls would have to retain and resist six to eight feet of earth. The greatest width of these outsize footings juts out to one side of the wall, under the earth. Sideways, or in section, it looks like a leg and a huge slab foot. If the dirt tries to push the wall down, it will have to pull up that foot with it. Anchoring walls to footings are upright rods with long L-bends, and also a "keyway," a groove formed by a bevel-edged 2 X 4 along the top of the footing, designed to catch the cement that fills the block.

We stuck other vertical reinforcement in two rows staggered 32" apart, one row outside, one row inside, spacing calculated by block and holes in block through which rods were to fit. Tom drove many stakes in the trench floors to mark the top level of the concrete-pour. He must have used the batterboard lines and story pole to establish the stakes, and a spirit level to level them with each other. I remember trying to clean out trenches again, from dirt and rubbish fallen in. They had to be kept clean, and I almost went mad trying to work and scrape along underneath the rebar and points of wire, reaching my hands between and under, to scrape and collect loose dirt with a little scrap of board. That was the day before the first pour.

Even in the later stages of building, with plenty of men to help, I could not hear the arrogant trumpeting roar of an approaching ready-mix truck without feeling my stomach sink. That dreadful roar which meant hard desperate work ahead. The overbearing, snarling motor, the overbearing yellow pomp of the slow-revolving barrel. Out the lofty blue cab hopped a stocky, cold, sullen driver. You have to treat a mixer driver like a varlet: shout and bellow at him, command him by small waves while your eyes are busy elsewhere. He always sends too much; it's always too dry at first; he doesn't cut the flow fast enough. In a rare perfect situation the thing is easy; the stuff runs like soup, you have nice straight channels and runs, and all you have to do is watch the concrete shoot and pour and fill your trench, and maybe swing the chute a little to change the direction of the pour. Most other times hell breaks loose, one way or another, utter smash and mess, and you have to break your back setting it right.

It was done in stages, as we ordered it. We never took more than two truckloads in a day. The pouring of all footings was spread out over time, and some were left for next year. Perhaps we should have done all ground and foundation work at the one time, rather than try to tie together the two segments later.

The first mixer carried seven cubic yards (3 X 3 X 3 feet), and we started at the southeast corner, under Sarah's (later Amy's) future room, and the south wall of the hall angle. The greenish gray stuff came down the chute and vomited into the trench, moving in around the rods, piling up, moving on, piling again so we had to pitch in with shovel and hoe to keep it moving. Heavy labor, heavy stuff. Vertical rods began to sag and fall under the press of it, so we had to rush around and stand them up again or they'd have been engulfed and lost. You push yourself to the limit in the sure knowledge that the cement is going to set hard and solid in an hour or two. So we worked as fast and hard as we could push ourselves, dragging, hoeing, shoveling the heaviest muck known to man, outside of adobe mud. Resistant. It does not want to move. It wants to stay. And every time I had to wade in up to my shins.

Before concrete sets, do the following: locate tops of marker stakes; remove excess concrete by shovel and wheelbarrow until the surface is roughly even with stakes; float surface until smooth and the gravel sinks below; go along with a marked stick and lift up all vertical rods until bottoms are in cement and tops are no higher than estimated height of block (or they'll have to be cut off later). By the time all this is accomplished you can barely pull up the rods, for the concrete has started to set. Tom had made a wooden float, a rough kind of trowel, out of a 1 X 4, with a whittled wooden handle.

Later in the morning we got another load, of eight yards. To let the truck inside the site and close to the north wall, I'd shoveled away a pile of dirt; and Tom had bent down rods already set in and had laid planks across the newly poured footing. But when the huge monster backed in, its wheels broke the planks and squashed the concrete. Eghh. Chingar.

By the end of that, footings were in and set for the east wing and partway down the north wing. I don't recall if we took more pour next day, or how we staged it. The basement might not have been ready yet, so we might have held off. The cellar-end was the roughest to do. We got another 15 yards in two hauls to run the kitchen and cellar. Even backed all the way up to the kitchen trench, and with 14 feet of chute, the shit couldn't be poured directly into the cellar; so Tom built two wooden chutes. They reached the edge of the hole, and we wet them with the hose.

Concrete came grating down the long truck-chute, into the first wooden chute, slowed down, then, pushed by more coming behind, oozed and shuffled down into the second wooden chute and finally dropped over the edge into a puddle below. Best we could do—drop it in a pile. Tom worked alone down there, distributing the pile to the forms. I couldn't help him because I was busy upstairs, finding stakes, hoeing, shoveling. I worked from a plank slung across the trench, trying to keep loose dirt from falling in on the cement. "This is a bitch," I heard Tom gasp from below. He must have been killing himself. Concrete doesn't want to stay fluid; it gets harder to move by the minute.

It was after this pour that we trudged over to Tom's car for lunch, and dropped into the seat so drained and exhausted we could barely move to open sandwich papers. I recall the tang of dill pickle, juicy, cool, sour.

Tom was a quiet man. He worked quietly, humming a little to himself but most of the time not a sound. You can't carry on a lot of talk and get work done, but even so he was very still. He loved to mind his own business. Dry-humored, he could make the others laugh. Liked to tease. Used to tell Ann he'd done something wrong and it was too late to fix it. Told her he'd ruined the front door, cut holes in the wrong place: "Too late now!" Knew he'd get howls out of her. He laughed with a wheeze, out of smoky windpipe and lungs. Perhaps because of his bad teeth, he would hide his smile—his mouth would work to resist smiling. "I'm gonna fire you," he would tell me when I goofed.

Here and there in the house can be found hammer dents, slips of a tool, which go to show that Tom wasn't quite as careful in his finish-work as he could have been. These marrings are scarcely noticeable except to a baleful eye trained like mine or Bartell's to see them. My eye knows what faults are there. It doesn't look for new ones. And my eye is beginning to see faults in context and diminish their importance.

It must have been hard for Tom in the beginning. I was supposed to be the supervisor, but I didn't know anything; so he quietly ran the job all the time it took me to catch up and understand what the job was about. It took me a while to learn enough just to keep watch and control over the doings. This is because for the first months I tied myself down to performing a laborer's job, instead of studying the plans, watching Tom, asking and learning. But I did get wise at last. Some time in July, with Tom's hearty agreement, I gave up the idea of working full-time, in favor of concentrating on plans. supervising and fetching materials.

I had begun with the vision of building the whole thing myself, with one or at most two helpers. But I let go of that idea. The house was built by myself and a crew of four, five, sometimes six men. I started Tom at $3.50 an hour, and raised it to $4; laborers I paid $1.50 with small raises the longer they stayed; block-masons, $2 or $2.50; adobe-layers, $2. Francisco Baca made $3 to $3.50 an hour. These wages were below union scale, but union scale is an empty phrase unless you work on a union-run job, which is very rare around here. Unions are not strong, and men won't join and pay high dues because unions never get enough work for them. I think I paid the highest non-union wages in Santa Fe. Most of the time I was making a payroll of about $450 a week.

I wish I'd known Spanish. Even in English Tom and I had trouble understanding each other's descriptions of things, or positions: did I mean upright? flat? did he mean edgewise? length? depth? height? Hard to convey exactly what we meant. And I wish I could have joined in the talk, and understood those wisecracks Tom was getting off, perhaps at my expense. I know a few words: *Palo*—wood, board, stick. *Carreta*—cart, wheelbarrow. *Troca. Vamos a comer*—let's eat. *Martillo*—hammer. *Ventana*—window. *Lonche*—lunch in Spanglish. *Viga*—any kind of supporting timber, joist or beam. *Pala*—shovel. *Regla*—ruler. *Plano*—level. *Cual*—which. *Este*—this. *Ese* - that. *Cimento. Bloque. Cuadrada*—square. *Clavo* nail. *Mescla*—mix. *Ya voy*—I'm coming. There is no *sh* sound in Spanish, only *ch*—therefore, shit is *chit*; shakes are *chicks*; shopping, *chopping*, short is *chort*. Piñon sap, hardened, is *trementina*, which used to be chewed like gum. English words are mixed with Spanish: *Si, catorce inches alla. Reech mortar, metal lath. No cheque*, it doesn't check.

Tom: round face, dark, with deep lines from nose to mouth; build, medium, but with round, stooped shoulders from so much bending, so much retiring, so much humbling and ducking. He was not powerful-looking, but he was as strong as anyone on the job. He had shaggy brows and sad, weary eyes with long lashes.

"I won't lie to you, Bill, I'm drunk." He said he'd never try to work when drunk. He got drunk about once a month in '64 and was out two or three days, delaying us. Next year, he made a great resolve and got in trouble once. He was arrested, drunk, with another drunk who was driving his car, put in jail and fined heavily by J.P. Albert Jenks. I paid to get him released—$136 including towing and storage of his car. Tom paid me back in two or three weeks— "working for the county" as he put it.

I could always tell he was drunk the minute he got out of his car. The weaving, lurching, the transformation was queer and horrible. I hated to see him drunk, it made my skin crawl, the loose grins and slurred talk and staggering motion. It always happened in the middle of the workweek, so we were held up and delayed, and it made me mad. He knew I didn't like it; but if I'd fired him it wouldn't have made any difference in his life, except he'd have been out of work again. He must have lost many another job because of drink, and must have known he was risking this one.

The first time it happened I was sure he was drunk on the job. I still think so. But that time, I spoke to him. I told him I didn't want him coming to work drunk. He took it as an insult, said he wasn't drunk but that he'd had something to drink the night before. He was hung-over, in other words. He offered to quit. I told him I didn't want him to quit. Next day, the mood was bad on both sides, and he came to me again and said he was going to quit. "Why?" I said in dismay. He said, "You don't seem to be happy with us, our work; you said I was drunk." So I ate crow. I swore I didn't know what I'd do without him. I'd be lost without him. He agreed to stay.

But he didn't stop getting drunk; and I didn't stop getting fed up, although I made no more comment.

I think in the end he liked me, respected me. I know I seemed strange, in my beard and horn-rim glasses and Old McDonald straw hat, and who was I? What was I? But we got along well next year. He liked Amy, and he and Ann were a funny sparring combination. "Mister Weatherford," she persisted in addressing him. She kept asking him how things looked, and weren't they beautiful; and out of perversity he would grow more and more laconic.

We left for later the remaining concrete work of areaway, steps and cellar floor. The base of the house was now established. Two or three times a day for several days I wet the new concrete with the hose, which was necessary to help it cure. The cement drank water like a sponge; weather was very dry and blowing all the time. With pick and crowbar, we pried off wooden forms, and Tom set the lines again and started projecting them onto the concrete footing for the block-work to follow. He used a chalkbox and line. Plumb under a corner, he would drive a nail into the concrete as something to secure the end of the chalkline. Then, using the plumb bob, establish another point. You powder the line well with chalk by running it through the special metal chalkbox, then hold the chalkline between two points, draw it up like a bowstring and let it snap its chalk on the surface of the footing. Thus you get a straight blue line to follow in laying the first course of blocks.

I decided to hire another man, a mason. Tom brought someone he knew to my house at Allendale Street, and we negotiated. Tony Olivas was a great stolid fat man, slow of body and brain, with very little English. He was amiable, had a fat chuckle, always laughed when I talked to him, a laugh of incomprehension. Long after his departure, I discovered that he drank on the job. Wine bottles lay everywhere. Under every bush I found another pint of

Italian Swiss Colony. He was stewed most of the time, I learned not just thick-headed, but sozzled. He was so heavy and slow he never showed it. Later that summer I heard he had died from a fall, probably died drunk. When he stood up on a plank, it sagged. Bob Kurth tells how on some job "Tiny" Olivas laid courses of block in line with the plank he stood upon—in a bow. "Too dry. Hah." "Too wet. Hah-ha." "That's all right. Hm." "No cheque, Tommy. Hm-hm." He had a pouchy face, like a huge J. Alfred Hitchcock.

Once I ran into a boulder when I was digging the piping trench from the well to the house, and I couldn't lift it out, nor could anyone else. Finally Tiny Olivas got down in the slit-trench, leaned down and by main force heaved out the rock. It was a narrow confine; I was sure his great hips and butt would get wedged in and we'd have to pry him out along with the boulder.

"Block" (singular and plural) is a shell made of pumice mixed with cement, a hollow shell with two large holes in it. Block comes various sizes and shapes, but we used "eights" (8 X 8 X 16) and "twelves" (8 X 12 X 16), along with some special corner, half and coping pieces. The corners come with smooth-cut ends for corners; most block comes with ends hollowed out or cut in a slot, for installation of steel windows. Too fragile for use as a masonry piece by itself, block must be filled full of concrete, so it becomes a form for concrete. Was it cheaper or better to do this, or to form and pour a solid concrete foundation? I assume the Bartells knew; anyway it's standard practice.

One rim of the block shell is thicker, wider than the other—more surface for mortar; therefore it is the top of the block, for laying purposes. There are various trowels. Olivas used a large pointed one. First he swept clean and dampened the surface of the footing. Then he troweled out a setting-bed of fresh wet mortar, roughly in the outline of the block shell, and laid up (not "down" but "up") the first block, which had to be a corner. Aligning it true to the chalkline, he laid a level across the top and tapped with the butt of the trowel until the block was level. I've since found that it's good to use plenty of wet mortar in the setting bed, and then press the block down level with the others. If you use too little, or it's dry, then you have to build it up or scrape it entirely off and begin again.

Block should be laid straight, plumb and level. Lots of checking, tapping, and nudging. Olivas mortared the end-faces of adjoining blocks. He picked up mortar with his trowel and wiped it off down the edges of the block as he held it in his hand. This he did in seconds, quick, easy and sure. Then he set the block down adjoining the first, with about a half inch of joint between. He picked off all excess mortar ooze with his trowel as he went along and flicked it back on the mortarboard. This was one of those unforgettable motions of masons, along with the tapping and leveling: the peeling-off, picking-up and slinging-off of surplus mortar. Brief, easy motions of a lifetime. He kept a can of water by him for moistening mortar and splashing the footing. He would put a splash on the mortarboard and scramble the mortar with his trowel.

We used various combinations to make up a 24" wall: three eights in parallel; one eight the long way, and two eights butted perpendicular to it; or a combination of eights and twelves. We would reverse the order on the second course, to make a stronger, interlocking bond between the layers. "Break bond each course," say the specs, meaning that joints of the first course, or "row" as the men said, should be bridged by the blocks of the next course; and so on up. To bring this about, you must start the second course with a half-block; and since you start at a house corner, the half will actually be the end of the full block that starts the second course of the other wall of the corner. Wall corners interlock like fingers, like dovetails in carpentry.

One picture equals a thousand words, a thicket of words, to describe what is so clear and sane, so beautifully simple and consequential. A child could see it and understand it.

The order of parallel eights is easiest to fill with concrete; with the other orders, the fill-holes close up smaller and smaller as the wall goes up. If the wall rose higher than three or four courses before it was filled, it was impossible to poke the concrete all the way down to the bottom.

Block walls had to be built up around the vertical rebar, which complicated the work; sometimes block had to be cut or broken to let the rod through. Also the architect specified that every other course of block be reinforced horizontally, full-length, by a truss-like wire frame called "Dura-wall."

Starting another course, Tiny usually held a block up to where it was to be laid before mortaring it, to check for size, position, etc.; then he would cut it, chop it or take up another shell as needed. He mortared the bed-block below, buttering it, slushing mortar all around the top outline of the shell, then set the new block down carefully in its bed, tapping and leveling. To align blocks vertically, he used the level as a straightedge. The work should be tight and flush up and down. If blocks are level across the top, then they must be plumb, since they are manufactured square.

Once the work rises above the first course and the chalkline, it must be aligned by other means. Tiny used a string and two scraps of wood that were notched out the width of a block, to straddle a block and hold a line tight along the line of the block. When all is done, walls ideally are straight, plumb, square, and neatly joined. Corners and intersecting walls are built up ahead so that the intersections are all tied in, like the corners. But in some places the masonry didn't come up to standard. Some places, like the two abutments at either end of the portal, the work was butted onto existing walls. Hope they don't crack, along with all the places where the second half of the house adjoins the first half. Olivas was a slob. The work got done, but was sloppy; and I know he cheated on his time.

The job was getting too much for me. I had to supply Olivas with mortar and block, also start filling block already laid. (The mortar mix was three to three-and-a-half shovels of sand to one shovel of Rich mortar, plus water.) So in the third week of May I hired Tiny's son Diego.

He was a broad-shouldered, strongly-built man about my age, with very erect carriage, a well-shaped head, full moustache, and spiky black hair growing up from the top of a high forehead. I think he fancied his own looks and build. He was wedge-shaped like a weight-lifter. And he was quick and ready of mind, and good enough at his trade that he could have gotten along as others do, he could have made out. But he was crazed by flaws. He said his back had been weakened by injury; he had a bleeding ulcer, a thirst for drink, and a yen to give up and die. He loafed at every opportunity. Twice he got sick on the job, threw up blood and had to go home. He loafed to the point where I chewed him out, the only time I had to tell a man off. When Diego got drunk, all that apparent upright manliness went down the pipe, and he was like carrion for crows.

He had a small son who spent most of his time with the grandparents. His wife was pregnant and was going to have to quit her job. So it was up to him. But he never tried. He made a stab at work, then gave up. Ann tried to help them and found out about the indigent hospital fund for him. By doctor's orders he should have been in hospital. But he wouldn't go. He was scared of hospitals. Some people can't be helped, they want to die.

I borrowed the Goads' electric cement-mixer and brought it to the site just in time, because it was hurting Diego's back to mix mortar in the mortar box. Tom had built the mortar box like a flatboat with a sheet metal bottom. The sides he cut out of 2 X 10 planks. It was 8 feet long, 3 feet wide. Until I got hold of the mixer, we'd been mixing mortar and concrete by hand in the box, and it was hard work. For one batch, 35 shovels of sand, 10 shovels of

Rich mortar; dry-mix them with the hoe; then start adding water, mixing and chopping with the hoe until you get a nice soft slithery mess, all pale gray. The blade of the mixing hoe has two holes in it to release some of the pressure and weight of the material, but it is still back-pulling, breathtaking work, rougher still with adobe mud, and murder with concrete, with gravel. Watery gravel drags against you, sucks against you. A 35-shovel batch in the box equaled about five light wheelbarrow loads. To mix in the wheelbarrow, combine one part Rich mortar and three-and-a-half parts sand, times two or three, to make a load. Tom could always guess how many wheelbarrow loads a given job would need. After a little experience I could begin to tell if a mix was too dry, too wet, too poor (not enough cement).

I squandered water all day long. Too much trouble to climb down the spider hole into the well and turn it off. Lengths and spaghettis of green plastic hose brought it out of the well to where we wanted it, and it spilled out all day long, clear, clean, tasteless and cold. God, it was good. I gulped it down while it ran all over my mouth, and I closed my eyes. Even after a bellyful it was hard to keep from going back 15 minutes later. When Tom Weatherford took a drink, he tossed down the hose with a watery, rasplng "Ahh." Pure, cold, never-ending agua. Like school children singing the same song on and on, voices clear and cold, clean and aimless and endless— you listen a while, then ignore it; but it's always there, a unison of high voices tracing and retracing forever a line of song. "Buffalo Gals." Colorless, shining and pure, the water flowed out of the well, more than we could ever drink or use in building, more than soil or trees could drink. Luxurious waste!

Hot weather was upon us and the work was not getting any easier. On the site there was nothing to break the glare that beat up from the ground, and the sun was so fierce and heavy it weighed on me like a huge extra burden.

The mixer rolled all day. Diego mixed mortar and wheeled

it to his father; I mixed concrete to fill the block. Six shovels of 3/4" gravel, six shovels of sand, two heaping shovels of Portland cement, with plenty of water chucked in ahead so the stuff wouldn't stick, all of it heaved one, two, three into the mouth of the revolving cone-shaped can of the mixer. Then I took a little water, sloshed a little in on top, watched, waited, dripped in a little more water, waited, then grabbed the lever and heaved the can over and dumped it into the waiting wheelbarrow.

Power for the mixer came in two or three extension cords from the meter pole, 120 feet away. Tom tapped into this arrangement to run the "Wizard." Clokey, the driver from Kauffman's where I ordered all my sand and gravel, tipped me off to different grades of sand: "concrete sand," which was coarse, and "plaster sand," which was finer, costlier. So we had two sand piles going, one fine, one coarse, for block-filling.

Clokey was 65 or 70 years old and used a hearing aid. Tall and bent, under a stiff cap with a visor, he had a seamed, humorous, droop-nosed face. His mouth was wide and mobile, with a grim expression that spread out wide in a smile. He had a deaf, high-pitched, squawky voice. When he delivered sand, he was never in a rush to get away again but would always jump down from the cab to have a word with me or the men and look at the work. He would stand close, mutter in my ear, nudge me and burst into a cackle of laughter. He was abrupt; when his say was said, he turned and left. "Well, thanks a lot, Mithter Gateth," he called as he pulled himself up into the seat of the dump truck. He lisped because of missing teeth, or loose dentures.

June, typical: I drove the flat square shovel into the gravel pile. Beyond me, bent over a burro, Tom set the power saw to the edge of some lumber and squeezed the trigger. "Shheee ..." with a series of screams the blade sheared off sections of two-by for header blocks, and they dropped to a growing pile under the end of the sawhorse. Behind me, a newly-hired laborer, Leroy Anaya, was distributing mortar to the boards of Tiny and Diego, who was now laying block for $2.50 an hour like his father.

The mixer can clanked around empty. I poured a bucket of water into the mouth of the can and it sloshed around. I threw in the first shovel of gravel. The can rattled and crackled the gravel around, "aggregate, aggregate." More shovels: "crash ... crash ... crash," it racketed as it turned the wet stones. So loud I barely heard the screams of the saw anymore. Wet mortar in the mixer made a "Flump, flump," as it turned; concrete crashed. The mixer and can rotated at a certain slow, deliberate rpm, and the noise followed that beat. Revolving crashes. In went a shovel of sand, then another. The racket was dried and muffled by the sand, until the water soaked it. A bit more water, now it began to crash again, not as hard as before, but wetter, heavier, thicker. Plow, plow, each blade inside the can plowed into the heavy mix and the mix poured over the blades and slapped the can, with a wet, revolving, plowing sound and an oozy slap. Two sounds in the beat—"shadump, shadump." Gray drooled out the mouth. I had to jack up the can to a higher angle with a scrap of wood. Well-used and abused mixer: the rough crusty hide of the legs and can were so caked with years of cement that you could only knock a flake or two off with a hammer. I pressed a stick to the lip of the can, letting it wipe new drool off itself as it turned.

I checked the cement supply. The sack was empty. Tatters of sacks lay all over the ground, and dark gray spills of Portland powder. A hardened hunk of slobber lay under the mixer-mouth, and the mixer feet were gradually being buried under a drift of sand and gravel. The hose ran through the mess, a snaky stream running to the road, down the edge of the road, around a curve to a puddle in a low spot.

I hefted another 94-pound sack off the cement stockpile, swung it around and thumped it down. I punched open its belly with the shovel blade: one long slit up the middle, two slits crosswise at the ends. I peeled aside the layered paper. Dark gray powder soft as talc it was, 94 pounds packed tight. I tossed the shovel back into the sand pile.

Wheelbarrow waited nose-down, hunched and pointed into the other side of the mixer, its wheel down in a shallow pit. I grasped the lever, grabbed my little wedge as I lifted, and quickly and strongly heaved the loaded turning can over backwards and down until its mouth was inside the rim of the wheelbarrow. Most of the mix rushed out; some stayed. I let the can revolve as much of the mix as it could into the wheelbarrow, then I banged the lip of the can up and down against it, to shake out the rest. Then I returned the can, replacing the wedge, and poured another pail of water in and left it turning. I walked around, took the handles of the wheelbarrow, pulled it back, and turned and wheeled it out. My arms pulled down, locked and spread by the handles. I leaned my whole body to control the wheelbarrow: forward to drive it, backward to slow it, way backward to ease it down an incline. The wheel hit a clod bang-on, and part of the load slopped out the front. I backed up, shifted feet, turned my body and the wheel, and bypassed the clod. Down any steep slope, I bent my legs and crouched to lower the back end of the barrow, and let the barrow run down and roll to a stop. Meanwhile the load shifted, swayed and slopped with every phase of the trip. My eyes kept watch over the load to see that it was balanced in the wheelbarrow.

At the work point, the destination, I set the canova over the block wall. The canova was a box-shaped funnel that Tom made out of lumber, with sloping sides four to five feet long, eight inches deep, and sized to fit along the top of a course of block. Blocks at this point were four courses deep—32"—and a wheelbarrow load would just fill two holes. I shoveled it in, and when the canova was choked full, I took a long stick or piece of rebar and a small pointing-trowel, and prodded concrete down through the holes. I scraped up excess and slung it in. Then I shoveled more into the canova. Slowly the holes filled as I punched the stuff down through with a rod. It was hard to get the last bits of concrete out of the barrow, the shovel kept bumping into the rivets. Bit by bit I worked my way down the walls, filling block. I did this kind of work all day for many days: filling block, pouring steps, pouring the bond beam on walls after they were up.

My heavy work tools—the pick, the square shovel, the two pointed shovels, the mixing hoe, the barrows—are now worn but still stout and good after all the work. The shovels have no points left; the steel is bitten and worn. The pick-head has been ground to a round nub half its original length. The long shovel handles are worn, cracked, and ribbed with rising grain. (I remember how the steel shovel blades rang with a high tone when I pitched gravel out of them, or struck something with them.) The wheelbarrows, used hard, are still in good working order. They have a single wheel with balloon tire that never popped or went flat on me. The wheel is suspended on an axle from two strong 2 X 2 oaken shafts, which are the frame also for the large, deep blue-painted steel tub, and for the two legs. The shafts are rounded into handles at the back. These wheelbarrows are really built to last—$27 each, manufactured in Harrisburg, Pennsylvania. The wheels are packed with bearings and have grease nipples for the axles. The tub holds as much as a man can carry. We always scraped and washed out the tubs and tried to keep them from rusting and corroding. Same with other tools. Diego said once you had to keep tools clean or "by the end of the job you won't have no tools."

Diego worked bareheaded. His high bare forehead turned red, red-brown, then brown in the sun. Leroy Anaya wore a straw hat like mine and overalls. Leroy was 60 or more, handsome, easy in talk and laugh; he showed me the *trementina*, how to chew piñon gum. He told how he'd been a cowboy, riding herd up in the Pecos mountains in the 1930s. But Leroy was no stayer. He quit on me after four weeks. One day he said he had to go in town "on business." Next day he didn't show up. I heard he got a county job, a soft job riding around in a truck cleaning out culverts. He was Diego's father-in-law. One time or another I took on five relatives of Olivas by blood or marriage; Bonifacio Mendoza was the only soldier in the bunch.

The more men I took on, the lonelier I grew among them, shut out by language and other things, perhaps some mistrust. Alone together, Tom and I talked a lot. Aside from shop-talk, he revealed a good deal about his life. How as a youth in 1937 he'd made adobes in Tesuque for the Ernest Knee house that later became the Porters' house, and how he'd work like a dog all day and then carouse all night, and how those days were gone. How the war caught up with him as a New Mexico National Guardsman (already enlisted) and took him to the European Theater. He was in a half-track, a tank-destroyer, and went through North Africa and Italy. He recognized my ghurka knife; he'd seen the troops who carried them. When the Munros rode by on horseback, he recognized Pat Munro's accent as British; he'd been posted to Britain before shipping out to Africa. The war showed him and others like him a world they never would have seen otherwise. Tom noticed things, stored them away and retained them. "Goums" he remembered, the native French African troops who would be turned loose on a village to plunder it. He and the soldiers in his outfit camped in Italy for months after the war's end before they were shipped home. I think the war was the biggest event in his life; it dwarfed everything else, before or since.

The two of us had things to talk about. But that changed when I hired other men. I continued to eat with them, but it was lonely. June and July were hardest on me. Loneliness, plus terrible hard word and heat. Set apart and leery of workmen, I used to balk inside at getting up and going out to the site in the morning. Home was a haven, a rest of brief duration. I was so tired and craving of shower and rest. Some nights I couldn't hold my eyes open past nine.

One whole day I did nothing but backfill. Shovel by shovel the great heaps of dirt thrown up by 'dozer and backhoe and were pitched back into place around the foundations. Wherever the inside wood floors were going to be below ground level, I painted the block walls with cold liquid asphalt for damp-proofing. Then it was heave ho with the shovel. Ant-like, I learned something of persistence and patience. I was an ant man, with a mountain of dirt to move, and I moved it. I moved several mountains. Don't hurry, don't strain, just dig the shovel in, pull it back, lift, swing it on hip or thigh, and heave. Or slide the shovelful, drag it sliding on the ground and shoot it in the hole. Keep at it. Don't think how much, or stand and stare. I found my own pace. When I find my own pace in a thing, I usually discover soon that if I try to go against it or rush it, I start to puff and blow, I start to hurt. A man's own pace is part of his grain, his spirit. Supported and carried along by my own pace, I could work on through ferocious heat, without a break, and at day's end I'd be surprised by the lot I'd done. Outside our future north bedroom window there was a mammoth pile. Shovel by shovel, it went down and was no more. Another great pile of white powdery dirt stood outside the future east windows, and shovel by shovel, it went down.

I saw I'd restored things as they had been, the natural grades and slopes of the land. It was satisfying to clear off the dirt right down to the needles and cones and twigs. It didn't take so long. It never does. A man and a shovel can move a lot of dirt, dig a big hole, clear away mountains of trash, and it does not take as long as it would seem. Learn of the ant. Ant, beaver, working animals get things done by repeating the same small, dogged motions over and over at their own pace. I see how the Chinese can dike a great river with baskets of dirt. No job is too big. It can be done. If you just bend your will to it, keep at it, don't rush, little by little.

I don't like pick and shovel work, but I got a lot of it done. When I finally got a tentative plumbing layout, I had to dig trenches for sewer lines and the trench for the water supply from the well. The sewer trench led out of the children's bath, under the hall and out—and it was a bitch, so deep it passed underneath the footings. I had to work down between the joists and then burrow down below the hall-footing, for the retaining wall, and then across the hall and down under the two-foot-wide footing of the outer wall.

Pipeline trenches had to pitch at ¼" a foot. I checked this as I dug—¼" a foot comes to one inch in four feet, two inches in eight feet. I used a straight, eight-foot length of 2 X 4 as a guide. Placing one end on dirt, you level the board by resting a spirit level along the top; then you check the other end for its height off the dirt. It should be two inches above dirt over a length of eight feet.

The sewer line from the children's bath was to join the main line from kitchen to septic tank. Granted, this was crude work, it still had to be close, closely figured. I used the plumbing plan and I saw that I'd have to start calculating at the top of the main line, in the kitchen. Floor elevation there was 97', 6" inches, according to the drawings. I think these elevation figures work up or down from a base of 100'. The pipe would run at least six inches under that, or below the brick and sand floor; adding four inches for the diameter of the pipe itself, we're down 10", or elevation 96', 8". At a scale of ¼" a foot, I measured on the drawing out to the branch - about 48 feet. Then I applied the ¼" a foot pitch formula over 48 feet and came down to elevation 95', 8" at the branch. Working back along the branch to the children's bath, and rising now at ¼" a foot, over a distance of about 40 feet, I figured the elevation as 96', 6" at the start of the trench. This was four feet below finish floor elevation of 100', 6"; so I knew I had to dig down to a depth four feet below elevation 100', 6". The joists were set by then, so I could make up the finish floor elevation by laying a 2 X 4 flat across them, and therewith check the depth of my digging by measuring down from the top of that 2 X 4.

The dirt turned up red clay and very hard; it would barely submit to a pick. I dug down to depth on each side of the huge outer footing, and then grubbed a hole through underneath the concrete. I used Tom's big bar, five feet of solid steel, very heavy, with a chopping point at the end. I used it like a battering ram, slinging it with both hands, punching out the dirt crumb by crumb, then reaching in and scooping loosened dirt out with my hands. I gave a whoop when the bar poked through to daylight. "Finally?" Tom called with a smile. My arms and shoulders were weak from ramming that bar. I had to do the same under the footing of the master bedroom in order to let in the waterline from the well. I dug a deep slit to the well. It must have been four feet deep, and I kept cursing myself for making it too narrow to work in. Always dig plenty; never dig too narrow! That was another bugger of a job. There I ran into the boulder that Olivas pulled out, the only stone of any size in all my digging, for a mercy.

Some days it was all I could do to get in the truck and drive home. Nobody takes breaks in the construction trades. My men never took a break. So I couldn't either. Steady on, no sitting down, no breaks. I had to get used to never pausing, to being miles from any refuge or comfort except the cold running water.

Now for the floor, piers, girder and joists. As support for wood floors, Tom had to stake out locations of concrete piers. Center to center distances from pier to pier, and from wall to center of girder, were marked on the plan. In all, eleven eight-inch-square concrete posts were to be formed to support the girder, a long built-up wooden beam like a backbone running the length of the east wing. The girder was to be the central bearing for the floor joists. I dug square holes, about 16" by 16", for pier footings, and Tom staked them to control the depth. Depth and height of piers could be controlled from the side walls, by means of a straight board or joist slung across the gap. By then the wall plates, the joist-bearing sills, must have been set in place. These consisted of a continuous strip of flat 2 X 4 bolted the length of the block walls, to act as rests, or "bearing," for the joists. The wall plate had to be set level at an elevation nine inches below finish floor. Bolts to fasten the plate had to be set in the wet concrete fill of the block, about four feet apart and with enough bolt sticking up to go through a flat two-by and take a washer and nut. These were called anchor bolts; they were one-half-inch in diameter and had a hook on the end. Tom drilled the two-bys to fit over the bolts, and installed the wall plate, checking for level, shimming with pieces of shingle where necessary. Once shimmed, we had to stuff cement under the plate and fill the crack. We sliced it in under the plate with the edge of the trowel. Then, after the plate was leveled and set, we secured it by running the washers and nuts on the bolts. If too little bolt was left showing, some wood might have to be chipped away to make room for the nut.

Once the plates were in, a joist could be thrown across to give an accurate guide to the height of piers. We knew the girder was going to be five-and-a-half-inches thick; so five-and-a-half inches down from the bottom of the joist was the top of the pier. The piers had not only to be centered in line, but also of exactly the same height, for even and level support of the girder. Tom made box forms eight inches square inside. Piers were to be reinforced and tied to footings with vertical dowels of 5/8" rebar. That extra eighth-inch seemed to make them almost impossible to bend or break, let alone to put a hook in them as plans specified.

I poured the eleven footing pads with the wheelbarrow.

While they were setting up, I stuck the rods straight down into them, deep enough to hold but not through into the dirt below. After the pads set, Tom staked the box forms in place around the rods, and then, using a joist as guide, nailed marks inside boxes to show maximum height of piers. Where the floor stepped down, a two-stage pier was built, a pier plus a foot-high wood post: this would be a division between rooms, the last joist of one and the first joist of the other would rest on this one pier, one joist a foot higher than the other.

We ran the piers out of the wheelbarrow, shoveling the boxes full up to the nails. Before the cement set, I inserted metal straps to nail to the girder. Straps should have been rigid iron; but we just used plumber's tape, a thin flexible strip of metal, not strong enough. But it doesn't matter; that girder will never move.

The foundation end-walls had to be chopped out in notches to bear the ends of the girder. The girder, or "stringer," was built up of three 2 X 6's pieced end to end and nailed together, with solid spans from pier to pier, and joints bearing only on piers. Tom built it so that the joints in the three runs of boards overlapped, yet at the same time would bear on piers. He spiked It together with 16-penny, 20d or heavier nails. The lumber was West Coast fir, Douglas, much stronger than pine, straighter and better cured. "Weyerhaeuser," every piece was stamped with that small triangular tree-form trademark, every piece tinted an unnatural pink, perhaps a preservative. Natural color of fir is white, like pine. You can distinguish fir by the grain, the sheen, the tight knots, the hard surface. It's a hard material to cut and heavy to handle. The lumber of the joists was "native fir"—local stuff, not Douglas fir.

All this floor-framing work was tricky, I can see by my tangle of writing. Best advice for an amateur: level, relevel, check and crosscheck for level every time you turn around; also check and cross-check your distances. If you don't catch mistakes on the spot you'll have to straighten them out later and tear out finished work. A floor must be level, strong, tight; its underpinnings are what make it.

The work had to be done over a period of time, according to concrete setting-time. Weatherford ripped off the forms while the piers were still green—too soon, though they held their shape. Once the piers were set, the two-stage girder was placed over the piers from end to end of the east wing, and the straps were nailed to it. The lumber was not laid flat-side down, but edge-down. Same with joists. The strength of lumber is through the edge-grain. Edges of boards are the bearing sides, they carry the weight.

Weatherford marked off 16" centers for joists along the wall plates. At each window opening, the floor had to encroach deeply on the wall, so he had to calculate and adjust joist-length for that. Vents to air out the subfloor and crawlspace area had to be boxed or cased through the wall, through a pumice block laid flat so that its two big holes could serve as breather holes. Floor-frame under windows had to be headered out to size because widths didn't match the joist-spacing. To make floor-frame for anything extra or odd-dimensioned like a window, small joists are cut and set and tied to the main framework by headers, short cross-members that tie one joist or beam to another.

The floor plan came into scrutiny now because there had to be double joists under all frame partition-walls, and a triple joist at the first joist of the lower elevation. Extra joists were to provide extra bearing for partitions. The double joists were spaced to the width of the partition. Of course, any other than wood-frame partitions had to be supported on concrete.

By this time I had stockpiled about seventy-five 2 X 8's sixteen feet long. Though the joists were to be 14'8", including four inches of bearing on each end, I had to get sixteens because lumberyards only sold in lengths of multiples of two: 12, 14, 16, etc. The leftover scraps could be used for headers and other things.

Here let me mention one of the puzzles of building: the difference between the nominal and the actual, the dimension called for and the dimension actually supplied. A "two-by-four" piece of lumber is not two inches by four inches, or anywhere near it. It is actually one and five-eighths inches by three-and-a-half inches. This is how it rolls out of the mills, and it adheres to those actual dimensions without exception, so that you can plan on it. The only real 2 X 4 is a rough-sawn piece. The usual stuff, as sold just about everywhere you request it, is what is left after the mill has planed down the rough-sawn stock.

This holds true in all lumber dimensions. A "two-by-eight" is really one-and-five-eighths inch by seven-and-a-half inches, and a "one-by-six" is really three-quarters inch by five-and-a-half inches. Architects use the actual dimensions in drawing plans and sizing things; but in their specs or any writing, they refer to the lumber in its nominal, title size. Ambivalence and deception worm their way into this supposedly hard-as-nails profession. More later on that.

Tom laid joists on edge on the marked centers, toenailing them (driving nails in at a steep angle) into plates and girder. Then he fitted headers into the spaces between the ends of joists, and nailed the headers joist to joist. Midway between walls and girder, he fastened two rows of X-bridging, lengths of 1 X 4 cut to fit between the bottom of one joist and the top of the next, and beveled across the ends to butt tight to the faces of joists. The "X" is formed by the second piece of bridging, fastened next to the first but diagonal to it, to opposite sides of the joist. In the end, two chains of these Xs ran the length of the floor. With headers and bridging tying the joists together, the framework became very tight and rigid. Tom always left the bottom ends of the bridging loose until later. The top had to be nailed before the subfloor covered it.

I have some photos I took one day in June after work. The men had gone home, and it was very quiet and beautiful on the site in a hot, slanting late sun. The view to the north showed a dark green hill in mid-distance, whose dark peak rose up in a gap formed amid the farther mountains. Overlapping peaks. I used to pause from work to feast my eyes on that view. Sometimes the hill would be dark in shadow, and the mountains beyond bright in light; sometimes the hill was lit, and the mountains in shadow; always, overlapping peaks of dark and light, slopes declining smoothly one beyond the other.

The photos show the building-up of the framework, how the members bear one another, tie together, cross and intersect; how the piers hold up the stringer; how the stringer carries the joists and how the joists carry the floor. In the picture a small patch of subfloor is laid at a 45-degree angle to the joists. Seen from one end, the joists seem like many even, calm waves or ripples, with the little bridging pieces riding over their backs. From the side, a powerful perspective forces itself on your eye. The eye races along the sunny edges of the foreshortened joists. Between them the bridging, sideways, seems limb-like, arms and legs. The effect is almost audible, like two tones repeated, tink tank tink tank. Repeated also, the long shadows cast by the sun. Music.

By May 18, I had 2,350 adobes stacked nearby, ready to go, the first of 20,000. Rather than tear the cover off our land, I bought the adobes from a man in town who hauled them out here by truck for a price of 13¢ apiece, $130 a thousand. In the '40s, Tom said they sold for 5¢ or 6¢ apiece. Between 1964 and '65, Sixto Montoya raised his price ½¢ apiece, pleading the rising cost of labor.

Before long there may be no more adobe-makers. There's no living in it. Mike Barela died in harness last fall. Lucero is an old, old man and Montoya is in his sixties. No young man would think of making adobe for a living. It doesn't pay enough for the heavy hard work you have to do. Once you hire help, even one man at the lowest acceptable wage, the cost cuts so far into earnings that you can't make anything. For low wages you hire dregs; for a rough job at low wages you can only get dregs. So unless you slave all by yourself or put your sons to slave for you, it's a losing business. Montoya, who is the only man making adobes in quantity, will eventually lose, though his other business of sand, gravel and dirt-hauling may take up the slack. Barela worked until he dropped. Lucero will do the same—still strong enough at 70 and older to make adobe. Small, wizened old men, still making forty-pounders by the thousand, one after the other, by hand. They can only be made by hand.

Contractors don't use adobe any more. Builders use block or wood frame; it's so much cheaper and quicker. Block costs 25¢ apiece, but block will build an 8-inch or 12-inch wall much faster, being twice as high as adobe—8 inches versus four inches. Also, a length of 16 inches versus a length of 14 inches. It is lighter and easier to handle, mass-produced and uniform, and makes those straight, plum, sharp-cut walls that builders and home-buyers seem to favor these days. Wood-frame construction is cheapest of all, can be insulated, plastered and weather-proofed, and faked to look like adobe in outline. Block can be left unplastered; adobe must be protected.

How do you make an adobe? First: good dirt, clay dirt, which is almost everywhere, the ground you walk on, New Mexico terra firma; take a handful and squeeze it; if it holds together in the squeezed shape, it is clay enough. The more clay the better, it can be cut with sand. Dig under the grass. Grass and roots, as well as twigs, stones of more than pebble-size, and debris should be discarded. The purer the dirt the better. When you have a good heap of dirt piled, hoe some aside for mixing. Add some sand, not too much. If adobe is too clay, it will be brittle and will crack and break; if too sandy, it will crumble. Add plenty of water, make a puddle, and then start chopping. Mud must be thoroughly hoed and mixed. Buford said that straw does no good, but the Indians and the Spanish all scatter in some straw. So, might as well mix in a very light amount of straw to propitiate the local demons. Straw should be broken in bits, scattered lightly. Too much straw makes a straw adobe, very weak, just as weak as a sandy or stony adobe. If adobe is mixed too dry, it will be weak. Dirt must be thoroughly drenched, soaked, hoed over and over again.

Build forms, frames containing two or more rectangles 4" by 10" by 14". That's the standard adobe size, though I've heard that in the old days they made them 12" by 18". Make the form of two-by lumber, remembering that a 2 X 4 will give you only three-and-a-half inches of thickness, which is the usual commercial thickness. If you want four inches, you have to rip it out of something larger. You could get a form made of sheet metal.

Set the form flat on clean level ground, shovel in the mix and pack it until the form is filled to the brim. Smooth off the tops of the sections with a float until surfaces are sealed. Wait until mud will hold shape, then lift off the form. Leave bricks to set dry in the sun for a day or two, then turn them up on edge like pieces of toast. They should further dry and cure for four to six weeks, the longer the better. From settling of mud, the top faces of adobes always appear slightly concave, with a worked surface from strokes of the float. The bottoms are flat, perhaps a shade convex, from pressing against the ground. Edges are smooth as the boards of the frame. Flecks of straw appear, and sometimes stones, even bottles—you never know till you start using them. Their dust rubs off on hands and clothes. They rub and chafe each other in the stacks; there's always a pool of dust where they have been. Buford went around tapping them with a piece of metal, saying that if they gave a certain clink it meant they'd been well cured. I don't know. I never made an adobe, and the mystery of a good adobe is something to be approached through trial and error. A good adobe must be very hard, neither crumbly nor brittle, but very dry, strong and consistent, packed tightly and free of flaws, stones, grass, or too much sand. Adobes can look good, feel good, but be no good. You find out when you start laying them, cutting them, or when the first big rain falls.

I never saw a 'dobe that could withstand the channeling of water. Over time, the rain will evenly spot and peck adobes and gradually wear them and expose the inner pebbles; but a stream of water cuts adobe like a saw. Wherever rain runoff gathers and focuses, it will channel adobe; and poor adobes will be gutted and melted away. Local people have notions about adobe like notions about weather, superstitions that are sometimes right.

It is heavy work, loading and unloading, wheeling and laying adobes. Sixto Montoya was a short round man with glasses and gray hair; strong and beefy-fat he was. He breathed hard. He had a beaming smile of false teeth. His pants drooped below his belly. He wore a small fedora. "Adobes? I gonna make you a good deal." He had a grating, heavily puffing voice. I helped him unload, once. They passed the 'dobes as fast as they could handle them. It really sapped me, swinging those 40-pound cakes into the stack or into the hands of the next man in the chain we formed. Montoya sweated and puffed, working as hard and fast as the next man. No time for a breather, hubba hubba. If one dropped and broke, he replaced it. 'Dobes were stood on edge in a truck, never flat; otherwise they'd crack going over bumps.

To cover the stacks, I used roofing paper or sheet plastic that came in 24-foot-wide rolls. Hell to unwind and cut plastic, but it lasted longer than roofing paper. I can see shimmering, flapping plastic coverings, clear at first, discoloring to dirty milk with time. I can see it fluttering, hear it stirring lazily, gleaming under a dark storm cloud. I can hear thunder and wind coming. Plastic blew like a feather, was hard to manage in any breeze; it drove me wild, and finally I used only roof paper.

     Before starting walls, we floored the basement, laid the dining room floor, ran the areaway steps and built the areaway wall. We reinforced the slab floor with "welded wire fabric," a heavy wire net with six-inch openings, which we cut and pieced to fit the floor area, and laid in place on a graded, wet sandy base before the concrete was poured so that it would be enclosed. Around the floor-edge we set expansion joints, half-inch-thick strips of felt or fiberboard, to absorb expansion between the slab and the wall. Concrete expands with heat, contracts with cold, like metal. After the expansion joint was nailed around at the correct grade and elevation, and the wire mesh was laid in place, Tom made a drag to control and grade the concrete. The cellar floor had to slope out to the areaway for drainage. He established cellar floor elevation, and from that mark drew a line sloping toward the doorway and the area drain. At the corners, the floor could be picked up a little so the corners wouldn't trap water. Also planned was a small 26-inch-deep offset, meant to contain the hot water heater. The end of the cellar slope was the top of the floor drain in the areaway.

     With lines as guides, Tom staked (or nailed) long 2 X 4s along the side walls with the bottom edges lined up exactly with the lines of the floor. These were the rails for a drag, another long 2 X 4 with two small extensions nailed on top, to hold it on the rails. The drag would grade and control the height of the cement between the rails.

     The floor drain had to be set in before any of this could be done. The drain-line trench had to be dug under (or through) the area footings and straight out to "daylight," or through the earth and out the steep slope of "Echo Valley." It was about a 25-foot run. I fell to and dug the trench, which grew deeper, sandier and easier as I got up near the house. I started in the grove below the northwest corner and carefully worked around two big piñon roots that ran across my path: two arteries that I saved. Downtown, I bought a ten-foot length of three-inch-diameter cast-iron pipe, an elbow and a drain-cover, oakum and strings of lead. By close figuring we set all in place, with the top of the drain at correct elevation. We couldn't melt the lead, so we packed and tamped oakum and lead dry around two joints. I backfilled some just to hold the pipe in place. We installed this ten-foot length, drain, etc. because of necessity of concrete pouring; the rest of it I left to be completed by the plumber.

The block wall of the area was built up and filled, and joined to the cellar wall by corrugated wall-fasteners, small pieces of corrugated metal cemented in the block-joints, then bent up flat to abutting walls and nailed—a puny-seeming thing. All these deep-down walls were supposed to be asphalted outside against damp and seepage. The north cellar wall was constructed hollow, to save on block: two rows of eights, between them an eight-inch hollow, tied together by pieces of duro wall laid crosswise at frequent intervals. In addition, a concrete bond like a tabletop had to be poured over the top and run over the head of the door to form a lintel. This concrete was run from the same batch as the floor.

The truck backed up as far as it could and poured ready-mix into wheelbarrows manned by Diego and Leroy, who wheeled it over to a chute that ran over the edge of the cellar wall. I remember Leroy coming with a load and upending the wheelbarrow in one quick heave. Very strong motion, his hands up straight, thrusting the handles, standing firm behind the barrow. The gray stuff belched out all at once, perhaps half getting into the chute, the rest spilling to ground. Had to scrape it up with a shovel when I got the chance.

Tom had built forms to contain the bond beam, door lintel and the lintel of the two-foot-square opening that was to give a crawlway under the dining room floor. (We did not remember to leave a similar access through the other wall to the future living room crawlspace—ouch!) After sufficient ready-mix was dumped in the cellar where Olivas and Weatherford were slaving to spread it around and grade it with the drag, we ran the lintel, and hell it was. I waded out into it to get the farthest reaches. Shovel, hoe and feet snagged in the rebars, which were hidden under shin-deep ooze. I don't suppose I thought much of it, but I was on top of a wall with a nine-foot drop to either side, and the *footing* was tricky, no pun intended, rather both meanings true. I had to work like hell and watch my step both. Tom had told me mixer trucks would charge overtime beyond 45 minutes. All the while, the driver stood there watching us work our balls off.

Anyway, it got done, slab-slob. I worked the concrete in along the top of the wall, closing the top like a lid; worked it into the lintel of the cellar door, where the level dropped and the pouring deepened to a foot; worked it around the corner and into the lintel of the access. The cellar stank like a zoo from all the sweat down there. Tom and Tiny worked their way backward toward the doorway, dragging and floating the stuff as they went. (I'd sealed off the drain with paper to keep cement out.) We may have run some of the stair with ready-mix, too; or did we run it with our mixer afterward? Don't recall.

The steps were formed with bulkheads as described above. Small footing trenches were dug at top and bottom, and rods, three number fours, were run down through and cross-tied by short pieces every 18" or so. This cement had to be smooth-finished like the cellar floor. After the concrete had set a while, firm enough to walk on, Tom took paddles that he'd made, flat boards with handgrips on the back, and padded out over the damp cellar floor with a square steel trowel. He used several paddles, placing them before him to walk upon, and picking them up behind him. When in position, he kneeled on the paddles, leaned a hand on one, and troweled with the other hand. He took some dry Portland with him to scatter around in case he needed more surface "bonding," and also a can of water to scatter more wetness. He smoothed the surface over and over in sweeping arcs, working his way over the whole floor on the paddles. I think he pulled out the drag rails at this stage and smoothed over any ruptures. He covered marks of paddles and tracks before leaving any area out of reach. Normally, you keep dampening new cement surfaces, to cure them; here the hose would have pocked the surface; so on Tom's advice, after the cement hardened, I scattered about an inch of sand over the new floor in the evening. This would keep it damp and protect it.

The east cellar wall, a retaining wall 12 inches thick (the west wall is 16 inches), was raised to a height to provide bearing for floor joists. The overall span was 18 feet, but with a bearing in between, the spans were nine and eight feet; the joists were pieced to those spans and joined over the bearing wall. Plus two runs of X-bridging. Over this Tom laid a subfloor for a work platform. For subflooring, I bought random-length 1 X 8 number three Common—cheaper pine lumber with lots of knots, good enough for subfloor.

He laid it diagonally across the joists, cut and fit it to join ends over the joists. After marking for length, he drew a line from the mark across the board at a 45-degree angle with his combination square. He drove the last nail toenail through the edge of a board, to force it tight against the next. He laid 216 square feet very fast.

The trick was to waste as little lumber as possible, but not to waste time trying not to waste material. A good carpenter wastes little material and no time. By experience, good workmen like Tom and Francisco Baca know what methods or combination of materials will waste the least in the long run. "You have to think ahead," Francisco said; usually he had everything figured out weeks ahead, way ahead of me. Waste is hard to avoid because of the commercial lumber setup: for an 18-foot span you have to buy 20 feet and waste

16 inches, unless you can use it for headers or blocking. At this date, I have a big heap of firewood, mostly short blocks, scraps of two-by that couldn't be used; I burned a lot of it that winter.

It took about a month to work the walls up to the bottom of the bond beam, and about six weeks more until we were ready for roofing.

Adobe work went slow because of cutting the angles of the deep reveals at window and door. I laid a few adobes, but Tiny Olivas and then Diego laid most of them. Even taking into account the cutting, it went so slow with Tiny alone that I was driven to hiring Diego at $2 an hour, and taking on another man.

Same checking, et cetera, applied here as in blockwork. 'Dobes were kept level with each other and plumb; a line was pegged into a joint to guide straightness. The 'dobes were laid up in an alternate bonding pattern each course, to make up 24 inches of wall. We used mostly the dirt of the place for mortar. It was mixed in the mortar box with enough sand to make the mud workable. At the start of a day, the men would be idle without mud, so the mixer would have to work like hell to get out the first batch. He would have to keep at it to keep up, screening dirt, mixing mud, wheeling and delivering mud and adobes. Usually the mix-master would leave a batch soaking overnight to start the next morning. One laborer could keep up with one adobe-layer; but it took two to keep up with two; and with three layers, two laborers would really have to hump it.

In the second week of July, I told Diego to bring out a man he knew, to help us. His name was Bonifacio Mendoza. He was about 20, very handsome, very quiet, reserved and shy as a deer. He had a moustache and a goatee, and a slightly aquiline nose. He spoke little English in his high voice; spoke very little at all, even in Spanish. Those first weeks with us, when we ate outside, he ate on one knee as if he were triggered to run. He didn't talk with the others at lunch and he ate off to one side, apart, on one knee. I'd never seen anyone like him. He was a true countryman in an ideal sense, strong, speechless, hardworking, simple. He came from a "ranch," I learned long after, probably a little place with a sheep, chickens and a horse or two, at Trujillo, New Mexico, a small dot somewhere in the vastness. He was related to the Olivas; his cousin was married to a sister of Diego's wife. Not a large man, a little smaller than I, but a very strong and a hard, dependable worker. After a while, the others called him "Bonny" for short.

The workmen ate cold canned food, out of the can: Van Camp's beans and "Beanieweenies," cheap Maine sardines, Vienna sausages, canned fruit, Nehi and soda pop for drinks. Once in a while a raw chile for garnish. Someone brought a cooked trout once. Tom Weatherford liked hard-boiled eggs and scattered bits of eggshell on the ground. I had to keep after them to clean up their scrap and garbage. They all ate in fifteen minutes, were always finished with cigarettes lit before I was ever through eating.

Adobe mortaring is not so careful as block mortaring. The bed is shoveled on, dumped on by bucket, and roughly spread about ¾ of an inch deep or more. Adobes are turned over and laid topside down in the mortar, perhaps because the rough, floated, slightly concave top surface of the brick will dig in and hold better in the mortar. The ooze is trimmed off along the joints and slapped into vertical joints, which should be filled up. Use lots of mud. It's fun to peel off and pick up mud on trowel and slap it into holes and joints. It's fun to spread and work the mud.

When the courses reached floor level, Tom inserted 2 X 4 nailing-blocks between adobes at four- to five-foot intervals and at baseboard height, just above finish floor. These horizontal nailers were for nailing the baseboard, and had to be built in. As the walls rose higher, he prepared other nailers for insertion, anchors for windows and doorjambs and partitions. For windows and doors, he took a short scrap of 2 X 6 or 2 X 8, nailed it flat to a longer, wider piece of 1-by, which he tapered at one end down to the width of the two-by. Overall length of the piece was 16 to 18 inches. Then he stuck many nails into the top at different angles, so it looked like a pincushion. This block was inserted flat and full-length into the wall so that only the narrow end fronted the window or doorjamb; the rest of it would be buried and locked into the wall by mud slushed and packed in around those odd nails. Three anchor blocks were normal for each side of a door. He tried to keep the nailing faces of the blocks plumb with each other.

Rough frames ("bucks") could be installed later after all head and jamb anchors were set, or they could be set from the start, braced plumb and square, anchored and walled in as work progresses. I think the latter would be better because it ensures a close fit between wall and frame; however, frames must be continually checked and set square and plumb, along two lines, front to back and side to side. It would require more care and time, but in the end the frame would not only be square and plumb but also anchored and butted tight to the walls. Weatherford wanted to go the easier route, to save himself bother and save me time.

So the frames were installed later, and in every case came to a loose, shabby fit between walls and wood, with gaps left. Shims, pieces of shingle, were wedged in everywhere to force frames into square. I think the windows are firmly enough fastened to walls. But because only plaster covers the gaps I doubt that they're as draft-proof as they should be; and I am uneasy about doors, which slam and depend heavily on their jambs. If he'd made a really snug fit between rough jambs and walls, we'd be better off.

We needed to know window sizes very early on, but Kitty did not supply them. By then we had a plan from her with window centers marked and the distances center-to-center between windows.

I had to find out from her what casements were to be used, get the Andersen catalog and find the sash dimensions. After that, Tom could add the sash width plus the thickness of the two finish jambs plus the two rough jambs, to get the outside, rough opening. He marked the centers and measured out half that opening to each side to give a guide to the adobe-man. Height of window off floor was the height of the heater ("convector"), plus a bit, so adobe could be laid roughly up to there. To give depth into wall for the heater enclosure, the adobe wall was thinned down to 8" or 10". Adobe man had to keep sides of rough opening plumb with the nailers, and then to flare the walls back from the opening. Ann and I decided on a big bevel, much bigger than Kitty's drawing. The bigger the window, the bigger the bevel, anywhere from 6" to 8" on each side. To shape these reveals, 'dobes had to be chopped with a hatchet. Difficult work. Imperfections and rain washouts formed the ultimate line of the reveals. The finish work, the plastering, was done along those lines. They curve and wave to the side like curtains, a more living line than if they were straight up and down and square as a box. Kitty planned a small three-inch flare, and here is one of the numerous changes we made on the job and could make because I was the contractor. How different the house would have turned out if I'd hired a contractor! Changes in plan and detail would have been difficult or impossible. It would have been another house, not so nice in many ways.

If you take care you can chop a hunk off an adobe pretty close to where you want. Girdle it. Start denting it along the desired line, dent it all the way around, keep denting deeper until it breaks along the line. It's slow, and dust and grit spurt into your eyes and mouth. Halves and smaller fragments can always be used somewhere, so they are worth saving.

With plumbing plans and specs finally in hand by late June, I went to Cartwright's, to Chambers, and to Willie Sena to get bids for the job. I called Roy Butler but he was busy. I wish he could have done the work.

Cartwright's bid was almost $8,000. Chambers dallied so long I skipped him. Sena bid $6,973. After checking the references he provided, I took him on.

Tall, full-framed, with motions sure, contained, paid-out rather than wasted, Willie Sena, "The Plumber with a Conscience," walked with a measured catlike tread, and he inspired calm and trust. He was in his 40s, Tom's age, but his hair was all gray around the sides. He had a good, pleasant face, more flesh than lean, and his voice was deep, even, peaceful-sounding. He wore a narrow-brimmed gray felt hat, a khaki shirt, jeans and calf-length boots that laced. He spoke almost without accent, some double negatives and errors, but still comparatively well. He seemed so intelligent, so sure of his work that he impressed me and I believed I was in good hands.

He brought out his crew in July to get some of the rough-in done, the below-floor piping, sewer lines and vent stacks; and for several days the Sena team showed hustle. I had agreed to handle the digging myself, and I had to move fast. There was a flurry of activity and then the plumbers were gone.

Kitchen and portal fireplaces, back to back, had to be built into the walls, had to rise with the walls. Olivas was not capable of handling the work, and so through Tom's offices I hired another man just to build fireplaces, a man named Francisco Baca.

It's been hard to keep him out of the account until now. Every time I think of the building he edges into view. Somewhere, in the foreground, or to one side, in the corner of my eye, I see him standing there waiting to be introduced. He does demand attention, he insists on it. Francisco, come forth.

He posed for the camera by his masterwork, the fireplace. He put one long arm out and rested his hand on it, put the other hand on his hip and confronted the camera. He did not smile, but his brown eyes, so quick and shrewd, were unblinking and burning with pride. His face and his whole big frame were stiff with pride.

His speech was a strange combination of Anglo drawl and Spanish accent, of malapropisms and blunders. Spanish was his tongue but he spoke English easily, unselfconsciously, with a kind of native slow pace and twang and twist as if he'd learned it early from a Southwestern Anglo. "Moist," he said for dampness, wetness. "If the moist can't get out, it'll rot," he said.

"Yuh, Ol' Tiny kicked the bucket," he said, telling me of Tony Olivas' death later that year.

"M' bruther kin do it," "There ain't any cee-ment aun it." "We gotta put some cee-ment aun it, see." "After we get the angle-ahrn, then we kin make that whichever way you waunt. See." "But ontil you do that, you ain't gonna have no baund, see. Onless you gonna get some a that black goop." "Memorandum blade." "Masonary lime."

I try to spell his words and it's no good, I can't.

His laughter gurgled out like bubbles in a big jug. Or it was a queer small sucked-in laugh, a hiccup laugh.

It was necessary to keep up with him, if not ahead of him. Otherwise, he said, "I'm just gonna be standin' here lookin' at you. And you don't want that."

As soon as I met him and had a few words with him I knew I needed him. He sized up the job right away and fell to—no doubts, no delays. He was tall, a good six-foot-two, long-armed and fast-moving. He was in his 40s; a grayish stubble showed when he hadn't shaved, and there were a few grays among the black on his head. Always wore a cap, pushed way back on his head, or with the visor pulled way forward over his eyes. Long, rugged face, wide mouth with slightly jutting lower lip. A shrewd humorous face, quick, foxy as well as impatient. He wore dark green work shirts and faded dusty pants, wore blue denim overalls for inside plastering. Large hands with cracked skin along the forefinger, always dusty or crusted with dry mud. He hitched up his pants with the insides of his wrists; probably wore out work-clothes like a meat-grinder. Checking the level of a thing, he'd say, "Looka that. Right on the money."

Can do. There was something grand and rare about his confidence, his handiness. An awesome energy seemed to possess him; there seemed nothing he could not build or repair or cope with. He hated to be idle. He hated having to wait for materials. He was hard to assist because he had to do everything his way and all by himself. Referring to Mike Lujan's way of work, that blend of idleness and finicking, Francisco said, "Boy, it sure makes the day go slow." He said it with understanding, pity, contempt. Yet with him it was more than getting the day over with or the job done, it was a matter of keeping on the move, on the do, and with things like fireplaces, a matter above all of achieving a good thing. And he could never get enough praise. I discovered that he could make mistakes; but he was sensitive about his work, and he'd duck it by saying, "I thought you said that's the way you wanted it."

He said he used to be a welder and used to work with his brother, the hunch back, Cruz; but he couldn't stick to it because he didn't have the patience. His impatience was exasperating. A couple of times he drove Tom wild. They argued about something and finally Tom said, "Oh the hell with it," and walked away from him. That's as close as I ever saw Tom get to blowing up. Without really meaning to encroach, Baca took over the job sometimes, set himself up as my interpreter, in place of Weatherford. He couldn't wait, couldn't help it, but I know Tom resented it. Baca worried and chafed and stormed about certain carpentry that needed doing ahead of his plastering, and that was not done because Tom couldn't get hold of his brother's table saw. I passed this along to Tom at lunch and he fired up: "Oh he's crazy."

But Francisco made very few mistakes. His other faults—thin skin, tendency to evade, to bitch about all the difficulties he ran into in every job - are common to all of us. They don't dent my belief in him, my sense of him as a kind of demon or god, taller and superior to the generality of men, capable of anything on earth, propelled by superhuman nervous vitality. He would rather do everything by himself and not have to wait and depend on another man. He probably would have preferred to build the entire house himself. He could have done it—bitching, kicking at it, fighting it but in the end prevailing over it and achieving something good.

He looked something like Robert Ryan, and also something like Wendell Clark, my cousin Edgerley's father. I saw that from the first, and that may be why I felt peculiarly at home with him, and with his speech, with its odd native-Anglo turn. I basked in his assurance and his savage hard work. He took up the fireplace plans that baffled the rest of us and made sense of them. He knew what to do and where he was going. Nothing stumped him. He knew what materials were needed; he understood the smoke-shelf, how it should lean and project, and how to go about building it.

Before Francisco came, we had the foundations for both fireplaces built, one a quarter-circular pad in the kitchen corner, the other, two pads for the legs of the outside fireplace, then the block-work solid up to about hearth level. Block, adobe, firebrick, common brick, clay flue, welded angle-irons, lath and plaster, wood frame: all came together in these two complex adjoining structures. We didn't follow plans in every detail. Didn't use exact radial dimensions given. When Baca shaped the dome of the kitchen fireplace, I controlled the shaping by eye. (It's off-kilter, but that doesn't bother me much now.) He built up the face-brick all around the bottom and up, forming the rectangular opening and sides up to lintel height. As he came to the hearth he began laying the floor of it, of firebricks instead of the cement slab called for in plans, large pale-yellow bricks laid flat and mortared with fire clay, a very dark gray almost black stuff that looked like gunpowder. Then the inner walls: firebrick laid on edge. Walls had to lean inward as they rose. He must have braced them up while the cement set. I expect he laid up these inside walls before putting up the front or facing walls that form the opening. Also, I think he built up the portal fireplace step by step and back to back with the inside fireplace; they share the same chimney, through different flues.

I bought a length of 4 X 4 X 1/8-inch angle-iron. (Plans called for a bigger piece, to carry twice the bricks we actually put up.) I took the iron to Cruz Baca, Francisco's small, red-haired, hunchbacked brother, to be bent in the same curve as the radius of the fireplace, to serve as a lintel. Cruz snipped the iron in a couple of places, then heated it at his little forge, a brazier on a stand, a coal fire he fanned with a bellows. A funnel and hood and chimney stood over it—all homemade. It was a tiny shop and Cruz ran it alone; his house adjoined the forge. When the iron was hot he tonged it up, flung it over his anvil and hammered it. He had to keep heating and hammering the iron until it turned the bend.

This was the lintel upon which the brick dome would bear. Baca placed it flat across the opening, with the angle toward the back so that a row of brick could be laid along the iron and continue the brick facing he'd already started. This row of brick he laid on edge. From there up over the whole face of the dome, our pattern required cut bricks, halves, butts, etc. Red common bricks were hard to cut evenly, without shattering. To form the swell of the dome, the bricks had to be placed so that they would curve and converge upward and backward all at once. Baca devised a supporting framework of several lengths of rebar, which he bunched and wired at one end so they converged. The rods were bent in a curve.

Bunched at one end, splayed out at the other. They were ribs for the vault of the dome. He covered the ribbing with metal lath; then laid up mortar and bricks over the lath. Before the top was closed, he had to finish the smoke-shelf, the step-like projection formed inside the back wall, at the top. It closes the throat to a narrow passage at the juncture of fireplace and chimney. The top of the step was cement, shaped into a bowl.

This was for draft. Smoke and flame are drawn up the back wall, narrowed and forced into a reverse curve, an S-curve, before rising up the chimney: principle of forcing smoke to follow a certain shaped, upward course which draws it more strongly. Like water: the more you narrow the neck, the harder it pushes to get through. Francisco molded the top of the smoke-shelf, then coated the inside of the dome with cement.

Flue comes in two-foot oblong sections that weigh over 100 pounds. Flue shell is about an inch thick, of terra cotta. You build one section on top of another, and this is the lining of the chimney. The 12 X 12 flue of the kitchen had to bend as it went up, and Baca set the first section perpendicular, cut the end of the next on an angle with the "memorandum" blade (Carborundum), then cut the next to fit perpendicular to the second.

The portal fireplace is hard to explain without plans. We poured a six-inch reinforced bridge slab, then built low brick shelves on top of it, on each side of the hearth, which is simply a firebrick floor on top of the slab. The space under the slab was left empty for firewood storage. To form the arched opening over the hearth, I had welded a framework of angle-iron and channels (strips of thin metal formed in channels), from which to hang lath and plaster, and upon which to build brick and then stud framework. The framework was meant to carry roof beams of the future portal where they intersected the fireplace. Above and behind this angle-iron contraption, another one similar was placed to carry the structure of bricks and flue. No dampers were installed in either fireplace: bad mistake. Once in a great while there comes a downdraft and ashes fly all over the kitchen. Not to mention the heat-loss from inside.

Francisco cut adobes with his big pointed trowel, and his motions were twice as fast as anyone else's. Arm and trowel hacked away in a blur of motion, dust exploding into his squinting face. Once he had built in the fireplaces, built up the walls around them, I kept him on to help finish the other walls. The adobe work was dragging, wasn't getting done. We poured concrete lintels for doors and windows. Tom formed them with ¾-inch lumber for the sides and bottom. We made lintels long enough to bear six inches on each side. To keep these box forms from bursting, Tom nailed splints of wood across the top and also led wire through holes drilled in the sides; the wires were passed through and clamped on either side by scraps of wood. The form-bottoms were boards cleated together by two cross-pieces the width of the lintel. These cleats he beveled along the sides, and peppered the tops with nails, so that when the form was ripped off they would stay locked in the concrete to provide nailers for the window frame. The forms were held up in place by props made in the shape of utility poles or trees: a 2 X 4 perpendicular and another horizontal across the top, with diagonal braces. He put three of these gallows-trees at each opening, to support the form with its load of wet concrete. Lintels were two feet wide, eight inches thick, and reinforced by five parallel steel rods. All had to be poured by bucket, up on a scaffold.

Weeks before, Tom had built four-foot squares for scaffolds, square frames of 2 X 4 braced diagonally. The squares could be stood upright in pairs or threes, then braced together with boards, two on each side nailed diagonally across each other, from the bottom of one square to the top of the next. Then two-inch fir planks were thrown over the top of this framework, to walk upon. We started using scaffolds when the adobe rose past eye level, and from that point on it went slower and slower. Mud had to be delivered up to the scaffolds by bucket; adobes had to be heaved up.

Where cross-walls intersected main walls, adobes had to be locked and tied in, as previously described; intersections had to be started from the bottom up, not added later. There were only two adobe interior partitions, between kitchen and dining, and between living and guest; the rest were all frame, which doesn't look as good but saved much labor and time, a saving that seemed necessary then. Toward the end of July, Weatherford began to mark out on the subfloor the plan of partitions, which he proceeded to build as the walls rose to completion.

I hired another laborer to dig for the plumbers: Sam Pratt, a cowboy by aspiration. He took off every weekend, sometimes Fridays too, to ride Brahma bulls in rodeos. He was a nice-mannered boy of 19, straightforward and hard-working, though because of rodeoing, it was hard to hold him on the job very long. He worked six weeks in '64, four in '65. Narrow head, long face with a notched, jutting chin. Wore glasses, cowboy hat and shirt, and boots. He had a bad stutter. He claimed he won enough riding bulls to meet expenses, and he could also make good money horseshoeing, his main sideline. He labored for me as long as he could stand it—and needed the pay.

Sam was recommended to me by his brother Schofield, an artist and cabinetmaker who we hired to make several pieces, including a sideboard for the dining room. Schofield was skillful but arrogant and vain of himself. When the time came, he way over-charged us for the sideboard, and we learned that he'd been egged on in this by his parents and by people we had thought were friends. Schofield was spoiled and haughty; Sam was unassuming and straight.

By the end of July, I had six men on the payroll, $560 gross wages in a week. No later than ten after four on Friday, I dropped what

I was doing and went and sat in the truck to sweat out the payroll. From each man's gross earnings, I had to deduct Federal income tax and social security, 7¼% of the wage, half of it deducted from the paycheck, half paid in by me quarterly. Some paid heavy taxes. Baca's gross was $140, but he took home $117, because he had only two exemptions, so he paid over $16 in withholding. Weatherford earned $140 and took home $131, claiming no dependents. Bonifacio earned $60 and took home $51, with no dependents. Baca said he always counted on several weeks off during the winters, because he spent that much time working for the government. Depressing for a man to realize how much time he spends working for the government.

I did all my figuring first, in a little notebook, and then wrote off the checks all together. Had to move fast to get it done by 4:30. I paid them, said goodbye—"See you Monday, eh?"—and they got in their cars. Francisco was always first off in his pale green coupe, off in a cloud of dust, waving as he went by. Bonny got off pretty fast, too, in his snappy brown compact, with a wave and a flash of white teeth. Last of all, Tom in his rickety '56 Ford that gave him so much trouble. Muffler no good, it rumbled obscenely—wumble-wumble-wum-wum-wum, a kind of throbbing grunt. He waved with an almost imperceptible weary smile, and his car lumbered on around the curve; I heard it lumbering and groaning up one hill after the other. Then they were all out of earshot and it became magnificently still there.

I hung around, mooning over the job, feasting my eyes on views through empty openings. The country through those rough apertures seemed so close and sharp-focused, so marvelous, in a way that it never would after the sash and glass were interposed. I used to stand in our bedroom-to-be and just gaze all down through hall, kitchen, dining room and out the window at the far slope of the arroyo. Those slopes and folds looked so near, so alive, shadows behind bushes so distinct—you could see so much and there was nothing but sweet sunny air between you and there. I stared at the roofless rooms open to the blue sky. Wide open overhead, not even any beams or vigas then. Various frames would intercede and shrink these one-beyond-the-other openings, and it would never be the same again. I got another telescopic view standing against the far (west) dining wall and looking back through the house out the bedroom window. Nothing would compare to this. Somehow the windowless reveals and embrasures intensified a sense of the outside, trees, stirrings of air, the very texture of air and sunlight. I saw tiny insects flying in warm light. Wonder and beauty in the empty openings of a half-built house! I was reminded of the view, the spy, all through Edgerley house to Weeks's hill in sunlight on the far side of the road. But that wasn't nearly as fine as this.

I've forgotten to mention my truck. I couldn't have managed without it. I bought it in April before we started building, a 1960 Chevrolet half-ton pickup with a long bed (eight feet long by four feet) It was painted white and turquoise blue; it had turquoise hips, as Ann said. It cost $1,150, was in fair condition, and with Its extra capacity box has given faithful service. I've used it hard. It has hauled sand, gravel, trash, bricks, adobes; sometimes it almost foundered under the weight. I'll keep it on until it gives up, I guess. The floorboards in back are warping and buckling. I've abused it, never washed it. But that's one of the beauties of a thing like this— you don't need to keep fussing over its appearance.

II

Reduction in force: once the adobe walls were up, I let Francisco Baca and the Olivas go.

August was quiet. Tom and Bonifacio were not gabby; Sam talked a little with Bonny, teasing him, but it was quiet. We all worked steadily. It was a good time for me. We got along, nobody was shut out by Spanish, and there was no uneasiness. I never felt anyone was trying to get away with something, as I always did with Diego around. Bonny ate with Sam, I ate with Tom, sometimes we all ate together. Mostly deep blue days, white clouds, deep blue shadows moving over the hills, yellow daisies blossoming and tiny yellow snakeweed. The land was spotted with yellow.

Heavy rains once in a while—they'd started in July. One Saturday in July Bartell called me to find out if my adobe work was protected; it looked like a cloudburst coming from the east. I looked out the kitchen door at Allendale Street, and then I went for the truck. I raced out to Arroyo Hondo trying to beat the storm. A stupendous gray and black war of clouds was coming over the hills; the first drizzle wet the truck as I turned off the highway. It looked like floods of rain, a smashing rain. I bolted down the dirt road. The sky was still dry when I got to the site. Walls were all naked of cover. I took the one roll of roofing felt, 100 feet long, heaved it up and unrolled it along the top of the wall. I got one wall covered before the wind began to blow. I threw on planks, bricks, blocks, anything that came to hand. Somehow I got all the walls covered and rain held off.

Air currents over the base of the mountains were tricky. The mightiest threats would veer away less than a mile from us. Rain on the other side of the arroyo, dry here; rain in town, dry here. One afternoon there was a fearful sight, a gigantic dust cloud about 15 miles south, pale brown, boiling and coming on fast like an enveloping explosion. "That's sure gonna hit here," said Tom; but it never did.

Rain striking adobes at a certain slant and concentration melted them away like snow. The mixer switch would shock at a touch. Pools in the tarp would have to be heaved out. Lumber grew heavy, then went cupping and twisting as it dried. The subflooring cupped. Water beads ran into puddles on plastic. The air was clammy, redolent of piñon, suddenly chill like in the mountains. Sticky mud on everything. Men waited in cars a little while, and if the rain didn't quit they went home, not to return even if the sun shone 20 minutes after they left. Rain pecked away the surface of new concrete, brought red sand to the surface.

We ran the bond beam, the "ribbon." It took about three days' unrelenting drudgery. The Goads had taken back their mixer, so I rented one. Bonifacio mixed, Sam wheeled, Tom poured. Or I poured, Bonny wheeled, Sam mixed. Forms ran the length of all walls, wired and braced, with nails inside marking correct elevation; five runs of rebar all the way, to be lifted up inside concrete as we poured. From plans we could read how high off finish floor elevation the top of the bond beam was to be at any given location. Once finish floor was established, on subfloor, or by a mark on the block, we could mark correct concrete elevation on the forms by use of a story pole. Then, drive a nail through the board at the mark, so the point would stick through inside. Several nail points at intervals would give enough guidance.

Foot by foot the pouring advanced, as one man scooped out the barrow with the bucket and swung it up to the man on the scaffold, and the high man would dump the bucket inside the form. Bucket by bucket. My gloves got holes in them, and wet cement soaked them and seeped into little cuts and burned sorely. Around the northeast corner we came, and on to the deep drops at bath and kitchen; it took forever to fill in behind the bulkheads at the drops. We floated the surface as we went, trying to scrape off excess water. "Too poor," Bonifacio cried to me, relaying from Tom that I'd put too little cement in the previous batch.

Treadmill work. But at last it was done. What we had achieved was a thick, continuous two-foot-wide rigid frame that adhered to all adobe walls and pressed down on them with great weight. It was supposed to be eight inches thick, but that was impossible, since the adobes were not uniform in elevation; yet the ribbon-top had to be kept level at certain controlled heights. Our ribbon varies from six- to eight-inches deep. It is split-level along the south kitchen wall, one level to bear kitchen roof-beams, the other to bear the portal roof-beams. Some lintels were run concurrently, with extra-deep pouring and reinforcing, over the garage door, for instance. After some days the whole thing dried to ash gray and we pried off the forms and threw them to the ground.

Before bedroom subfloor and partitions could be put in, the retaining wall between the brick floor and the crawl space had to be built up higher. I used 4 X 8 solid pumice blocks. Bonny laid them in, on edge. We had a combination of pumice block and wood, a 2 X 4 plate, both to raise the wall and to extend the wood floor base to the edge of the brick. Then Sam and Bonny laid the subfloor over all the bedroom wing. Tom nailed down the partition base-plate, a 2 X 6 cut at doorways to the width of the door, plus finish frame, e.g. a 2'6" door, plus 1 and 5/8, plus 1 and 5/8. Positions of doorways and center-to-center dimensions were given on plans, so he measured them off on the subfloor using the scale of a foot. He used his chalkbox and line to mark out the lines of the partitions. I discovered that our bedroom door was too far to one side to give that view through the house I liked so much; so I had Tom stand a board upright to simulate a door frame, while I trotted down to the end of the dining room and had him adjust rightwards until I got the view I wanted aligned through the openings. And there he marked the floor for the doorway.

He built and raised the bearing partitions first, between the hall and bedrooms. To find the stud length, he may have taken the given height from finish floor to top of bearing (bond beam)—say eight feet, as in the two upper bedrooms—and then subtracted the double head-plate, 3½", and the difference, ¾", between finish floor and the top of the base-plate. Then he cut the 2 X 6 studs all exactly the same length for the one ceiling height and set them 16" on center down the line of the plate, with double studs each side of doorways. He assembled the frame on the floor: studs, doorways, double headers on edge for doorway heads, and the head-plate—a whole section at a time, up to an intersecting wall. Then we helped him raise the section and stand it up on the base-plate while he braced it upright and plumb. He toenailed each stud into the plate with two 8d nails each side. He used his level to check for plumb two ways, side and front. To brace the framework, he nailed a scrap block to the floor to anchor the brace and ran a board diagonal from that block to a stud.

Where other partitions intersected the main wall, he doubled up the studs and spaced them apart with blocks, to provide a good nailing base for the first stud of the intersecting wall. Thus the principal frame partition-walls were assembled and raised, section by section. Later, Tom would work out the smaller framing details and problems, such as the closets and the sections of the children's bath. We had to erect the bearing walls first so we could put up the roof timbers. After the sections of partition were up, he nailed a second head-plate on top of the first, sectioned so that it would overlap the joints of the first.

We proceeded to the vigas. Armijo, who ran the wood yard on Canyon Road, supplied vigas at 40¢ a foot. He was a gentle, soft-spoken man with a pencil moustache, a reasonable businessman compared to the miser Moya. Francisco called Moya "the Spanish Jew." He told how he went to Moya for vigas, and Moya tried to charge him 65¢ a foot. "I told him, 'I want 'em made out of wood, not gold.'"

Armijo's yard was piled with cedar posts, firewood and pine logs in various stages of finish. There were a great heap with the bark still on, and some peeled vigas, and a couple inside the shed with men astraddle working on them. I learned from the specs that viga logs may be "spring-cut" if cured for over a year, but at their best they should be "fall-cut"—I guess because the sap is not running anymore and they are through growing and bleeding. I tried to get "fall-cut" timbers. But what I didn't know and nobody tipped me to, is that vigas must be big in diameter to look good. If I had it to do over, I would not accept any 16-foot-long vigas less than 7" or 7½" in diameter at the heavy end. Taper should be slight, no more than an inch overall, or maybe 1½" in a very long pole. The heavier they are the better—they need to look heavy, and they look bad if they're skinny. I did well enough on the second half of the house, knowing this. But then I didn't know, and so accepted all Armijo sent, except one or two with bad checks (cracks) and sap. He sent some good ones; he wasn't trying to pull a fast deal on me. People accept any size pole as long as it's a "real viga;" so he was just supplying standard goods. But some of those long sticks are puny. And all the six-footers in the hall are no more than 5½" to 6" in diameter; they look scrawny, little picks, no good. I began to have qualms, but it was too late. They had to go up or they were already up.

And if I could do it over, I'd do all the staining myself with Minwax. Ann and Betty did a few and Bonny and Sam did most, using the poor-quality stain that I bought. The stain wasn't brushed down into the cracks enough, and it wasn't controlled and wiped carefully, and only one coat was applied.

Two of us carried vigas into the house but it took four men to hoist them up into place, laying one end up on the bearing, then lifting the other to where Tom could reach down and grab it. We spaced them no more than 29 to 30 inches apart, on center. There was a certain number of spaces to each room, determined by the length of the room. In a room 10 feet long (120"), three vigas would work at 30-inch centers, three vigas, four spaces. If ceiling material such as shakes were used, spacing was figured according to the minimum spans required by the shakes. I don't like vigas set too close together. Ours are spaced from 26 to 30 inches apart. Once the spacing was worked out, Tom marked centers on both bearing walls, rather, all three bearing walls—the hall partition made a third bearing. We rolled the vigas to their marks and he tacked boards across the tops to keep them set in place temporarily. (He did not adze the viga bottoms flat as specs directed.)

To level the tops of the vigas with each other—for a level roof-deck—he loosened the tacked bracing-boards and used them, leveled them, then blocked and shimmed vigas up to the boards. Then he tacked them again until they could be built in. We placed 31 sixteen-foot vigas full-span across all bedrooms and full-span across the laundry hall and the small rooms behind, the sewing, bath and dressing rooms.

Roof and ceiling drop one foot along the north wing, between the master bedroom and the kitchen; a nine-foot ceiling steps down to eight feet at the turn of the hall. To do this, to carry vigas at the lower level, we bridged the hall with an 8 X 8 timber, which was let into the adobe wall on one side, into the stud frame on the other. The studs had to be offset and capped with headers to support this square timber. The bottom of it was set in at a height of 7'4". It's a fine, dark-weathered timber, with a red hint in it, a wine-red hint.

I got all timbers for lintels in July, as they must be built in below ceiling-height, built into walls. Tom said to try Olson's Lumber Company, so there I went, not knowing what I was in for.

I drove through a tall gate on Agua Fría Street and found myself in an enormous, dilapidated place, part ruin and decay, a good twenty acres of mad accretion and stockpiling of material. The farther I drove, the more vastly it opened up, beyond and beyond. First thing through the gate was the ruin of a mill of some kind, maybe saw, and roofless, yawning adobe walls, and rusting machinery. Broad fields of hip-high weeds rose over and through old boards and scrap, and waved overall. The road turned past the crumblement of mill and down to the left, in the direction of the river and Alameda. It passed stacks and a honeycomb lumber-storage built along an adobe wall. Everywhere was scattered, jackstrawed, unknown and forgotten lumber. ("A million feet," he claimed he had.) The road led past towering stacks of boards toward a dominant high hum, the singing whine of a machine. On a clear day it could be heard all over town; it would follow you to the end of town. A building stood there, a planing and sizing mill inside a tall shed of Olson timberwork roofed with corrugated sheet metal. It was a peak roof, very steep, very tall. Behind this structure rose a great conical stack where the waste was burned. It was at least 30 feet high, sheathed in rusty sheet metal, with a screen over the top. The rest of the yard rambled downhill toward the river into remote reaches, backwaters of material, sheds, abandoned works, truck carcasses.

As I nosed the truck down the road, a man came alongside,

and I knew right away he was the lord of this fastness. When he straightened up he stood over six feet, with his cap brim dropped low over his eyes. He was in his 60s, blue-eyed, white-haired, hook-nosed. He had wide jaws, wide at the hinges, and had a small gruff mouth, down-turned at the corners. Lank, tall, no fat on him—he pitched in and wrestled lumber himself, was on the job all day. His helpers never did things right enough for him. He stood bare-chested under the bib and straps of overalls made of striped ticking. With long strides he moved about his domain of junk, poking here and there, chin up so he could see out from under the cap.

He studied my list of needs. "Eight b' eights," he said with emphasis. "Eight b' *tens*." He addressed not me but the list. His voice had a violent, gruff note, explosive. He ruminated aloud. He seemed and sounded spare, secretive, cagey, grave. No subjects to his sentences. Repetitions. He revolved hundreds of details in his mind; dealing with an occasional customer off the street only tickled the surface. A million feet. Taxes. Insurance. The family stronghold: compounds to build, labor no good. He had an overriding drive to pile up goods, stock, worth, and go on making more, day after day, all day long. Whether or not he sold or moved the stock didn't seem to worry him. He moved enough to make way for more.

I think he owned sawmills that supplied him. Mills were up near the mountains and the source of supply. I saw one up beyond Pecos. A power forklift carried logs from a pile to a deck in the shed of the mill. Hooks grappled the logs one by one onto a conveyor. A flatcar on rails rode beside the conveyor; a man on the flatcar controlled the grappling and moving of the log back and forth, into and out of the huge circular ripsaw. Another man sat in a controller cab facing the saw and controlled the sawing. The log must have been gripped by the flatcar somehow, because after the saw had ripped through the length of the log, the car raced the log back, edged it over and brought it forward head-on into the blade again.

The slice fell off sideways onto another conveyor, and further along a man guided boards along and culled waste, pieces too barky to use, and sent them down a chute to a dump below, where I think another belt moved them along to the incinerator stack. The waste would have made firewood or slab siding for sheds, but they burned it all up. Useable stock traveled on into another system of sawing, manned by one or two men, where it was further cut and ripped to length and width, then conveyed down and outside the building to be stacked. I suppose the equipment is preset and geared at the start of a day to cut one dimension all day, or all week; I doubt if they ever stop and readjust to different cuttings. Five or six men operate the mill. Bang, clatter, rumble, rip—the din is colossal, you can't even shout over it.

A mill such as that supplied Olson with full-dimension, rough-sawn lumber. A rough-sawn 2 X 4 is fully 2 inches by 4 inches. Olson's was the only place in town that sold rough-sawn lumber. He also planed it; his planing setup was geared to shave the rough down to the standard board-sizes. They go by even numbers—2, 4, 6, 8, 10, and 12, the widest. Grading rules and specifications are set by the lumber industry; a system of minimum standards is set for each grade—select or common, "B or better," "C-select," "No. 1 Common," "No. 2 Common," etc. "Select" means clear of knots and flaws—"clear stock." (Through this experience, I can tell at sight the length, width and thickness of just about any board.)

Olson had the only planing mill in town. He made the smooth lumber, and maybe tongued and grooved it too; he sold T and G. But his specialty was rough-sawn heavy pine timbers. Six-by-eights, 8 X 10s, 6 X 10s and bigger, anything you wanted he either had or could get from the sawmill; so he must have owned a mill or two, or several, for all I know.

Beautiful things were to be found in that yard. I loved it there. Every time I got a chance or had a reason to go back, my mouth watered; I just about slavered with anticipation. Beautiful wood. Singing wood.

Those big ripsaws in the mills must be about two feet in radius and powerfully driven. That great, steel-singing radius running so high it whistles and blurs around, it enters wood and slows in pitch as the teeth start to war with the grain and knots; it cuts along the grain and straight through the bent of the grain. It nags and moans like a dog worrying something. It drives on all the way, spitting, grinding, raging on through to the end and out, free in full savage spin, steel-ringing, steel-whistling. As it shears through the log, it leaves traces of its passage. They show up on the lumber you buy: inscribed arcs, one chasing another along the face of a board. Once in a while the backspin, the back-whirl, inscribes opposing arcs that cut across the others. In our ceilings, we have boards and beams with a delicate sawn relief of arcs and counter-arcs. The path and signature of the saw. It stays through time and weathering. Subtle gray sweeps and rainbows chased into the wood. Memorials of the hard fight of saw against grain.

Such things I found in Olson's yard. That first day as we hunted for timbers, Olson said, "Like 'em black?" with his gruff gunfire emphasis. I was noticing the weathered hunks he had there, and all at once I realized I could have things like that in my house.

We found one here, another there; he peered in a heap and started moving and snaking the big bonks around, and we found another. Huge dark-brown and gray bars of pine. He moved one, levered it, and the pile clunked and boomed; he was skilled at moving lumber of any size barehanded, single-handed; there's a knack to it, something like judo, using the opponent's bulk to throw him. I watched him laying long boards in a stack, rolling them edge over edge, shaking, rippling them so they unstuck and slapped into place; I watched him levering 8 X 8s, swinging and tossing them so they fell where he wanted.

"I'll go find some hombres," he said. He peered away with his chin up and strode off to collar two workmen to help load the timbers onto his large yellow forklift. He didn't say a word to them, only pointed to the timbers with a stick and waved the stick toward the forklift. He seemed to ignore people utterly. The men tentatively made for the piece they thought he meant, hesitating; he jabbed at it again with the stick and waved. No words. He drove the forklift with its load over to a saw to crosscut the timbers to length.

His men wrestled the wood onto a conveyor, a track of cylindrical rollers, and pushed each piece along into place for cutting. Olson marked them to length in his grave, magisterial way. He switched on the saw. Its drive belt slapped and shivered and ran faster and faster, and a round blade about a foot in radius whirled and gradually rose to a scream. The blade was mounted on a radial arm that swung forward and back. Olson reached for the handle and pulled the saw toward himself and into the wood, once—"ZIT"—halfway through; he let the saw swing back and pick up power and speed again, then pulled it forward—"ZIT"—and the eight-inch timber was cut through. He prodded the piece on down the rollers and made ready to cut another. A man loaded the hunks into my pickup, which began to sink grievously under the weight.

Olson always figured out a price on the spot and declared it after inscrutable jottings and mutterings, punctuated by: "Forty-four feet. Forty-four feet … Sixty feet. Sixty feet … " ("Feet" meant "board feet" or square feet.) "That's about a thirty-five dollar bill," he said. He never gave me a ticket or a statement with the price on it.

I had a beard in those days. He wanted to know if I was an artist. Later he figured I was some sort of oddball, artist or whatever, always scrounging around for "black" things; but he trusted me to poke and browse on my own for hours on end. Folk Art Museum people would come around looking for weathered wood; Olson wondered if I was connected with them, with the "Old Folks' Museum," as he called it.

My truck staggered away under the load of lintels.

Tom Weatherford figured on adzing them off, but I told him to put them up just as they were. Like joists or beams, lintels need a four-inch bearing on each side. A 6 X 8 was slung diagonally across the face of the kitchen chimney, with the purpose of supporting the kitchen roof-beams, whose bearing in the wall might be crowded out by the chimney. It turned out there was enough bearing, so the diagonal was not needed; but we put it up anyway. It's a fine piece. It crowded one of the lintels to the hall, so Tom cut the end of the lintel to admit the diagonal 6 X 8. He cut it wrong and had to glue a piece in to make it look right; it would have looked better without that botch, but anyway …

A bunch of 8 X 8s runs overhead through the French doorway into the portal from the hall. Four are twelve feet long; by extension they serve as the ceiling for the utility closet inside the adjacent doorway. The wall of the door-reveal is the wall of the closet, and the back wall of the closet is the wall of the hall inside, 14 inches thick at that point. At this lintel passage, Tom had a battle with a twisted timber.

When wood twists, it looks as if the log has been wrung by giant hands; the grain turns around in a long spiral as the wood grows. Fibers want to grow that way; and when the log is cut into lumber, the boards and even heavy timbers will go on wanting to twist. And they do. One end will be okay, square and plumb, but the other end will be turning over, twisting. It's impossible to set such an ornery beast tight against another. I bought two big clamps, pinch-clamps mounted on long pipes, with faucet handles to turn and tighten. Tom used them, interposing buffers between clamp and wood to protect the wood. The beams stayed clamped together 24 hours. No. He tried nailing them together. No. Nothing would bring them together; still they are not snug and side by side, like good boys, the way beams should be to look prim and proper.

For the various linteled passageways, I figured out combinations to make up the needed depth: 38 inches from dining to living, four 8 X 8s, plus one 6 X 8; dining room niche, an offset about 26 inches deep, one 8 X 10 and two 8 X 8s—which could also have been three 6 X 8s, plus one 8 X 8, et cetera. I see that lintels are mainly construction devices to hold up the weight of walls in progress, wet walls whose weight is great. But I'll bet you could take them out after everything was set and dry, and nothing would fall down. That's Alan Robertson's theory, and I think he's right. You need them to build on, and they are better left there, they look so good.

I went back to Olson's to order rough-sawn decking for the bedroom ceilings and also the 6 X 10s for the kitchen. I took Ann along to see the place. She got big eyes for weathered lumber. It was her idea—I think maybe her best, her greatest of many gifts to the house—to cover the kitchen with weathered boards. She told me later how Olson had talked of these odd people from the "Old Folks' Museum" coming to get weathered stuff for their displays. Olson got a kick out of Ann; she brought a grin to his face, and the next time I went there, he asked me where my "secretary" was. He called me "young feller." He warned me that rough boards never came exactly equal in width or thickness, and that I'd have to "fight it." "Want to fight it?" he asked. He was so right.

His truck came out in August to deliver twelve 6 X 10 monsters and 1,200 square feet of rough-sawn 1 X 10 boards. It was a flatbed truck that dumped. The driver tipped the bed and the load slid off the back onto the ground. Before he moved ahead and dropped the other end off, Tom and I threw long two-bys crossways on the ground so the load would come down on them and be blocked up off the dirt.

Those 6 X 10s were fresh-cut and the heaviest things a man ever tried to budge. "Heavy dude," Sam Pratt would say. "It's a heavy dude." Twenty feet long, including four feet of excess, which I had to pay for but had no use for. I still have the four-foot chunks lying around; Sarah steps up into a stirrup and mounts Hondo from them.

The 1 X 10 rough lumber came anywhere from 9 to 10½ inches wide and all gradations between. Tom had to fight it sure enough. But first the partitions between the bedrooms had to be built. A 2 X 4 stud wall between bath and bedroom was offset and inset, both to make space for a closet in the bedroom and space in the bath for the basins and cabinets. It goes out and in. The other bathroom wall is straight; partitions for closets in Michael's room on one side, and for toilet and bath separations on the other side, were built out from it. The partition between toilet and bath is 2 X 4, but should have been larger to allow for the four-inch vent pipe going up through it to the roof. The pipe was too big for the partition; it's barely under the plaster. Toilets are required to have four-inch vents; but on a quick inspection of the roof, I find I was supplied with only one, for the children's. Sena provided a 1½-inch vent for the master bath; Butler a two-inch vent for the guest bath. Shows you what happens when you don't double check every item. Maybe it's not that important, but it seems funny that Bartell didn't catch this, not to mention the state inspector.

Between Michael's and Sarah's rooms, and Sarah's and ours, we erected so-called "sound-resistant" partitions. They were built according to Bartell plans upon a 2 X 6 plate, so that there would be no contact between one side of the wall and the other, except at floor and ceiling. As instructed, we put up two rows of staggered 2 X 4 studs with the blocking (or headers, or "firestops") set in on edge so the blocks didn't touch the other row of studs. In short, two frame walls on one plate. Though it might have been better with a 2

X 8 plate, the system doesn't work well at all since the rooms share the same floor. Though it may cut some sound, it's nowhere near soundproof.

After the stud walls between rooms were raised to the level of the tops of the vigas, the men installed blocking in every wall. The two-by blocks were cut of same lumber as studs, to fit between studs, then were nailed horizontally between them. It's a bit difficult nailing them all in line, easier to stagger them different heights, so you can nail through the stud into the block. This work braces the studs apart and stiffens the whole framework—grab it, shake it with all your might and it doesn't move or tremble. Tom put one row of blocks in eight feet of height, two rows in nine feet.

We laid the vigas up over the hall. We departed from the plans at the large recessed turn or angle of the hall. Instead of another bridge-beam and vigas running at right angles and ceilings divided in two parts, I got the idea of turning the vigas around the corner like the spokes of a wheel. Visions played in my head, of spokes, rays, basketwork. Tom and I did it, and the vigas looked so great turning against the empty sky! They never looked so good after they were roofed over.

Sometimes I would walk down this hall and series of rooms that were beginning to take form, and I would visualize each child in her room or his room, living in the future, growing up, studying at a desk.

We stained the decking with walnut stain cut to about half strength with mineral spirits. We brushed it on one face of a board only and neglected to paint the edges, a mistake that would plague me later. Boards that were cupped would be nailed cup-down, so we stained the cupped side. These rough-sawn boards were still white, a few weeks out of the mill. They were covered with a rough splintery nap that made them all texture and surface vitality. This added to their natural water-current lines, curves and pool-circlings of grain, and pink and yellow streaks and casts. When stained, a wonderful variance appeared within the pigment: red-brown, cinnamon, burnt tawny, dark chocolate brown. The finish was soft, subtle, with no sheen, "flat." They looked soft as nappy cloth. The thinned stain let the natural cast of the wood dominate the stain. We used one coat only.

Laid up side by side over the darker vigas, those boards sing. They are not too dark or heavy, but light, warm, soft. I remember looking into a room when it was half covered, and getting my first feel of that brown rough-sawn decking. It was one of the great thrills of building. The first thought to hit me was that it was like a barn, the inside of a barn. Ann was pleased when she saw it. I remember we had a picnic with the Bartons in that half-covered room.

Thankfully, we found the wisdom to steer away from the decking proposed in the plans: 1 X 6 tongue-and-groove commercial decking with edges beveled to form a v-groove when laid up side by side. This monotonous stuff, common in many houses, was scheduled for our kitchen ceilings, bedrooms and others. It makes a lifeless ceiling, ugly to my sight. It puzzles me why so many use it; probably it's quicker and cheaper to install, and it has a uniform "neat" appearance.

With Bonifacio helping, Tom laid decking over the children's rooms very fast, 725 square feet, in a couple of days. He nailed three eight-penny ("8d"—from 8-pence, perhaps) nails through each board at each viga, not always tight to the viga but with an eye to keeping boards level with each other, so as to provide a level surface for the roofing material. He had to fight deviations in width and thickness. He showed me how boards had to match in width down an entire run, or trouble would start in the next run. He nailed cupped boards cup-down in the hope that the roofing weight would flatten them out. Twisted boards can be sprung true when nailed down tight. But bowed boards with a warped edge are bad to work with. If the edge-warp were not too bad, Tom could sometimes hammer it in tight against another board and spring out the warp. He drove the nail into the edge at a steep angle and then when the nail was nearly in, he hammered hard, slamming the edge to force the board over against the other, and force it straight.

Now it was time for the giant shingles, the red cedar shakes that were to form the ornamental ceiling of the master bedroom. Wood split by wedge or axe makes a rough, rich-textured ceiling. I've seen split pine, as well as split cedar poles and shakes. Cedar splits clean. Buford Bartell had suggested shingles, which are smooth-sawn both sides and paper-thin, and which he had designed for ceilings in a couple of his jobs. He took Ann and me to see one at Jacona Ranch near Pojoaque. Ann liked it, and though I was not crazy about it, we thought we would try it. Shingles were lapped together between the vigas to form a chevron pattern.

However, I couldn't find shingles long enough to work— our vigas were not spaced close enough together. I noticed shakes at Big Jo Lumber and wondered at the time if they would work. Then Buford dropped by and showed me a shake he'd picked up and suggested I use shakes; said he thought it would make a wonderful ceiling. It was two feet long, plenty long. Big Jo could obtain them, but it would take weeks and cost more. I phoned Albuquerque Lumber Company and found that they stocked shakes. Some houses in Albuquerque are roofed with them. Shakes are primarily a roof-covering, very common in California and the Pacific Northwest. I ordered 18 bundles, and a few days later I drove down to Albuquerque to pick them up.

Shakes come in large banded bundles like shingles, except much bigger, about three feet long, two feet wide and very heavy. They were made in Washington State, and they cost $6.68 a bundle, with tax. Stuffed in tight, the 18 bundles barely fit in my truck.

Back in Santa Fe, I broke one open to try and work out

the ceiling pattern on the ground. Ah! This was forsooth a greater thing than a shingle! Giant cards of cedar spilled out, some 12, even 14 inches wide, and two feet long. It was gorgeous, awesome, exciting material. I felt as if I'd opened a treasure chest. The shake felt light as a big fan, but it was still a substantial wedge of wood, 3/4" to 1" thick at the butt, tapering to 1/8" or so at the thin end. What a ceiling these things would make! The varied widths, the variegation of the split faces, dark red-brown, honey-brown, rhubarb, purple so dark it was almost black, dark stripes running through pale, and the creases along the fiber like knife-lines running exactly parallel, and deep shadow-grooves like carved furrows. The wood fiber twisted where knots pressed them out of line. The edges curved with the grain, and some had a sheen like satin. I leafed them together on the ground in different ways until I figured out the pattern. I could hardly wait to see them up. I was almost jumping up and down.

On the bond beam between the vigas, flush with the interior wall, Tom built up small bearings like tables to serve as nailing surfaces level with the tops of the vigas. Making sure to start from the correct side of the room—which depends on where you enter the room and see the ceiling the most—we would start the rows of shakes with the point of the chevron toward that wall, toward the man working them. In that way only can the end of the second shake and the back edge of the third shake be concealed behind the first. The trick is to conceal both ends and one edge of every shake and, at the same time, reveal as much of the face as possible. No nails, no ends, no corners show; only one edge and beautiful face show from below. Someone needs to stand below to guide the work as it goes along. If carefully done, the faces of the shakes will reveal themselves one beyond the other, woven into strong design in an inscrutable manner. Braided Vs—women's Vs—one hiding behind the other, overlappings beyond and beyond.

In the master bedroom, we started at the wrong wall and ran two rows in the wrong direction before I caught it. Tom placed each shake, moving it, adjusting it till it looked best from below, marked and sawed off the wild end. Wild ends, which are nailed to the vigas, must be trimmed off to make room along the same viga for the next row. Tom used two 6d coated nails to fasten shakes, thick ends, to vigas. He discarded shakes less than five inches wide as too narrow to work. It was a slow process, at least two weeks' work to cover bedroom, bath, dressing room and dining room—533 square feet overall.

"This is going to be the most beautiful house in Santa Fe," said Ann. "Mister Weatherford," she demanded, "How do you like your ceiling?"

"Oh, it's pretty," he allowed in a muted voice.

"Chicks," Francisco called them.

"Chicks chingar," Tom grumbled.

On and on the roof preparation went, from the end of July to the middle of September. Before the roofer could start, the deck had to be completed and the adobe parapets, "firewalls," had to be built up to flashing-height, about two rows above deck level. The whole house had to be smoothly and completely sheathed over and a "dry sheet," one layer of 15-pound roofing paper, laid overall. Canales (gutter-spouts), sewer vents, furnace chimney, ventilator openings, skylight openings, all had to be framed, cased and raised through the roof deck to a certain height above the finished roof.

We spent a whole afternoon building little stud tables between vigas along the hall, because with the planned zigzag pattern of aspen latias, the aspen would run into the wall and therefore would need nailers and support. Wasted effort, we later discarded zigzag pattern and ran the aspen straight. Because of the press of time, we deferred the latia work until later. To make this possible, we jacked up the roof deck high enough over vigas to allow the aspen to be inserted later. Kitty Bartell's idea was to lay 2 X 4s on edge along the tops of the vigas, and nail the plywood deck to the 2 X 4s. The deck could be roofed over while leaving space for the aspen, the latias, short small-diameter poles running from viga to viga. On frame walls that bore vigas, the vigas needed to be blocked apart, so we provided the same stud constructions described above, or, simpler, a 2 X 6 on edge. Good chance to use up scrap. Butts of vigas should be doused with Woodlife.

No plywood was needed over board decking. It bore weight and was level enough for the roofing paper. Except the rough-sawn. Its varying thicknesses made sharp, uneven joins, an uneven surface that had to be eased to keep it from cutting the roofing paper. We nailed scraps of sheet metal over bad joints to make a slope from one board to another. Cedar shakes cannot bear weight, so they had to be sheathed over with another plywood deck. Tom laid 2 X 4s flat over the shakes at the vigas, and nailed the ends down with 30d or 40d spikes. Shakes were inclined to hump up because of the laying method, so he also laid a flat 2 X 4 in between, down the middle, not so heavy that it would break through. When the plywood deck came down on top, it helped compress the shakes without breaking through them.

At the laundry hall and small rooms next to it, the 2 X 4s running edgewise on vigas to give clearance for latias matched in elevation the shakes-plus-2 X 4s, so that the panel decking went down smoothly there. Where the elevation dropped, between sewing and kitchen, Tom built upward extensions of the frame wall to act as a curb, to keep the drainage plan of the roof. There was a special roof-framing plan that showed which way the different areas of roof were to drain—to east, west or south, and never to north. Between kitchen and hall, over the 8 X 8 lintel, Tom continued the curb by building another stud wall, to keep that whole section draining out its assigned canales, not onto the kitchen roof, which had its own drainage scheme.

All the sheathing, patching and small framing jobs are too complicated to describe. There was a lot of such work. If I'd understood then what I do now, I'd have tried at all times to bear in mind certain overall considerations. All the roofing work was aimed to make the roof drain according to plan, section by section; bedrooms and hall out the east; utility, bath, et cetera, out the south; kitchen and dining out the west. Everything had to be covered, sealed and eased to keep the pumice insulation from tearing through and spilling into the rooms below. The angle, the crease, between roof and wall had to be eased. The roofer could form most of the "cant strip," the sloping of the angle, with pumice; some of it had to be built and installed by Tom. He cut runs of 1 X 4 and "forty-fived"—beveled—the edges, so it could be fitted along the angle.

I did not see it then, but now I understand the relative elevations of floors, ceilings and roof, and their relationship and balancing. For instance, the floor drops a foot along the east wing, but the ceilings and roof stay up at the same elevation. The ceiling and roof drop a foot along the north wing, but the floor stays the same until the kitchen. At the kitchen, floor drops two feet, bond beam drops one foot, ceiling drops a few inches and the roof a few inches. Here the roof starts another section, a long run to the western canale. These factors made for a ceiling height of 9'10", almost 10', from kitchen floor up to deck, and ceiling height of 9'8" in the dining room.

We lugged the 6 X 10 timbers to the kitchen, Bonny and Sam in front, Tom and I in the rear. We slipped a stick under each end, for handles to lift and carry these heavy dudes. We placed one end on the scaffold, and from there up to the top of the wall, and from there over to the partition, and from there over to the little beam across the chimney breast, and from there over to the south wall. Then the next timber. Nobody fell or got a finger mashed.

We lined up and spaced the beams on the walls. Then Tom framed the skylight opening. He cut and headered-off one of the beams, meaning he made two Ts out of the cut beam, and fastened the T-crossings to the beams that paralleled the cut one. He nailed arms on top to support the headers, while he made them fast with two ½" lag-screws at each joint. Lag-screws are huge screws a foot or 14 inches long, with four-sided or hexagonal heads that you turn with a wrench. He had fastened arms, which were extensions of two-by lumber, on top of the headers, to hold them up while he screwed them; but the stuff was so heavy one piece pulled loose and fell to the dirt below.

For the skylight framing, we followed the detailed drawings Kitty provided. We knew the sizes because I already had the "skydomes"—ordered in plenty of time all the way from Maine. They were opaque white bubbles, not clear, of special plastic formed to deflect hail, and were bound into metal frames that would fit over and screw into our boxes. For the kitchen, we had to make a square opening 37" by 37", centered inside the beams and cut through the decking. Two other skydomes, oblong not square, were framed up through the sewing room and the children's bath. Tom called them "dooms." The lumber of the boxes had to be wide enough—2 X 10 or 2 X 12—that the boxes would rise higher than the roofing.

I bought weathered lumber for the kitchen ceiling at Olson's. I found colored boards all widths and grays—smoke, blue, salmon, red, brown, yellow. It was water-stained, burned, weathered. Tom was very dubious. But as soon as he began to lay it across the straw-yellow 6 X 10s, row by row it began to sing. The straw and wheat tones of the timbers intensified the blueness in the grays. Smoky blues! I keep looking at it, even now. It's the same in the guest room: weathered boards over yellow and brown log vigas.

I scrounged and rummaged through Olson's yard all one morning to find a truckload. "That's about a thirty-five dollar bill," Olson declared after consulting the little gnome inside his head. He should have paid me to take it away. Except there is no price for this stuff, it's beyond price. Smoky, cloudy with flashes and streaks of light, fading glows of light, hazy cloudy color—winter skies look like the boards in our ceiling.

One time I saw Olson show off the strength of a piece of lumber by jumping up and down on it. He laid one end up on a pile, walked up it and sprang up and down on it. I watched that tall old man, gravely, triumphantly, bounding up and down on a plank with all his might. He knew wood at sight. He claimed some pine could be stronger than fir, depending what part of the tree it came from. Olson's Fort, it seemed to me, something out of old times. The tremendous gate slid shut on trolleys. In the office was a huge fireplace faced with cobbles, and a cluttered desk about eight feet long and four feet broad, a tick-tock pendulum clock and a black metal safe painted and rimmed with gold paint.

He kept long slivers of scrap built up in crisscross stacks, which he used as blocking for his stacks of boards. He would toss three or four of these slivers across at intervals, then place the next layer of boards on top, and so on, to separate and ventilate the layers. Rough-sawn boards would be white at first, then turn yellow, then brown; edges and top layer would change, showing flecks and burning, scorching of silver and gray. Weather singes them as they lie there for weeks, months, years.

There were heaps of timbers and jackstraw piles of junk, scrap, and boards all twisted and bowed like rockers. Out one way you'd find viga logs lying in thick beds of curls and shavings. Out another direction, stacks of adobes, thousands, years old. Olson planned a compound of houses for his family; the adobes were not for sale. By the look of them they'd sat waiting a long time. The road led downhill into remote stretches of his yard. We found there the three pillars of our portal, amid a scatter of longer logs. What a find: three just thick enough, nearly equal in size, and short enough so I didn't have to cut into a big log. Beautiful.

"Are they dry?" I asked Olson once about some 2 X 4s.

"Dry?" he said. "Dry? They're light. Light as a feather. They'll blow away if you don't watch out." He began to chortle madly. "Blow away in the wind! Yeh!"

One shed contained a pile of vigas, rough-peeled, the outer bark trimmed, dark red-purple inner bark still on. I got all vigas for the next phase of construction here. It was a rough shed, one of the few provided to keep material out of the weather. It had a post framework and a siding of pieces of board, scraps, like a patchwork. That patchwork was fantastic—ash-gray outside, inside all stained by leaks. It was stained by water and weather and age, dark red-brown, black-brown, with weird striping, white, yellow, green around the knots, eaten and channeled and eroded. It was only a crude shed, but I'd like to have those pieces of siding. It was a museum. I could go back and back to that place. I'd find myself cooking up reasons to go back. What will become of it when the old coot dies? Rummage sale! Maybe Rover will burn it down to collect the insurance his father has amassed.

Rover Olson worked with his father. He was tall, lanky, his face long and thin, not square and wide like his father's. Not the same grip and tenacity, softer mouth, deep-set eyes in a very long face. He dropped subjects of sentences like his father. He was long-armed, very strong. Courteous. He waited on me once. Looking over some 8 X 8s, he turned one over and said, "Looks like a pretty good stick."

He told me they didn't carry cedar posts but that "Veeheel" might have them. He meant John Vigil, who used to sell posts.

Hidden down in the weeds I found seven good sticks and

true, for our dining room beams, 6 X 8s, all solid, straight and very dark. Seasoned. They were 18 feet long. Olson wouldn't sell part, so I had to buy all, though I only needed 12 feet to span the room. I used some of the leftovers for lintels; what happened to the rest I can't recall. I had more than seven remainders because Olson's man cut one wrong, then cut me another and hid the wrong-cut pieces in my truck—gave them to me to cover himself. Olson was out of sight.

They are dingy pieces—I should have scrubbed them. Before they could be mounted, Tom had to build a framework to bear them over the lintels of the sideboard niche and the lintels of the living room doorway. He also provided "furring," a framework for lath and plaster, where there was no solid wall, but a gap needed to be covered. This is an intricate wall, turns and returns, insets and offsets, to create opposing recesses, in dining and living rooms. The passage walls are deep, 40 inches, an illusion of thickness. They are really returns that form niches on each side. Both recesses are linteled over and furred up to the ceiling, lathed and plastered.

Before we wove the shake ceiling, I Woodlifed the beams end to end. With brush and can, I straddled the beams on a board. Feet dangling, I painted up to arm's reach, moved board, self, can and brush, and painted some more. It was unnecessary, but I like what the Woodlife, that evil-smelling, stinging poison, was bringing out in the wood, the red and yellow veins.

Bonny and Sam built up some of the firewall, and Francisco Baca did the rest. Before roofing, the walls had to rise above the deck to a height related to the depth of the pumice insulation, which would be a minimum of two inches and graded up from there at ¼" a foot. Thus on a 40-foot run, the pumice will go 12 inches deep at the high end, and adobes will have to rise 16 inches above the deck, or four courses. On a 20-foot run, such as over the bedroom wing, the pumice deepens to seven inches, plus about three inches for the cant-strip where roof meets wall, and that means at least ten inches of parapet above the deck, or three courses of adobe. The adobe is built up just high enough to take the pumice and allow the roofer to bind the roofing into the wall, as a shield against damp infiltrating through cracks in plaster.

The plumbers came, extended the vents upward and left flashing on pipes to be worked in by the roofers. The "flashing" was a collar of soft lead around the pipe, with wide skirts to be folded in between the layers of the roof and caulked with asphalt putty at the collar.

Tom built the canales of two-by, exactly according to Kitty's drawing. Fine work, one of Kitty's best details and Tom's best jobs. I took the woodwork to G&G Sheet Metal to be lined. They did a good job, and charged $6.75 each for lining the troughs and covering all weather-surfaces with sheet metal. The spouts were to be installed no more than two inches above deck-level, or so that the bottom of the canale is level with the deck. The back end of the canale is nailed or bolted into lumber so as to lock it into the wall, so that heavy ice loads won't rip it loose. You should be able to stand on the end of a canale after it is set. We didn't bolt them into the bond beam as plans indicated, but they are secure—three feet long, with one foot in the wall. They should be fastened to the roof frame, never to adobe. Nothing fastened to adobe is secure, unless with built-in anchors.

The roof-jack of the furnace chimney, which rose from the northwest corner of the cellar, through the dining room behind a frame partition, was a device left by the plumbers to extend the chimney opening up through the roofing material. A roof-jack was like the skylight boxes; one came with each ventilator also. The ventilator jacks consisted of metal hoods containing light tin flaps on springs just strong enough to draw the flaps shut, just light enough for the small fan inside to blow them open. Trouble is, the wind can suck them open, and on blowy days there is a constant irritating rap and tap of the vent-flaps. Roof-jacks are bound into the roof with paper and asphalt mastic. All such protrusions are potential leak-points.

I submitted the job to two roofers for bids. Farwell got it. He came out to look over the works. He was a heavy-set man, 45 or 50 years old, dressed in khakis. He had a tiny black moustache like Oliver Hardy. His speech was nasal and Midwestern, and he went on with many ruminative "mmm's" and "aaa's" between words. At his suggestion, I altered the plan to put another canale off the kitchen roof. Although one was called for in the plans, we did not build an adobe firewall between kitchen and dining room roofs, with a scupper through it for drainage from one to the other. Farwell didn't like "scoopers" as he called them, and said I'd be much better off with no firewall, and an unobstructed drainage run. So he inspected the roof and set a date for us.

Though all basic work was done, we had to hustle to be ready on time, had to clean everything off and lay on the dry sheet, called "dry" because no hot asphalt is used to gum the bands of roofing paper together. We worked one Saturday morning to get it done, because Farwell's men were coming Monday. There was much rain at that time, mid-September. I was always looking over my shoulder at oncoming clouds while I worked. Rains poured down through the deck; there were so many leaks it rained inside. Water found passage between deck and walls and ran down the walls; I watched adobe run down in gobs. Mudstreams piled up on the subfloor. After drying, the mud-dribbles still ran down the walls like congealed candle drips. Rain stained the ceilings in kitchen and bedroom.

There was nowhere for the water to go, no way to protect the walls against the knives of runoff. The three of us worked fast, hoisting the heavy black rolls to the roof, ripping open stickers and rolling out fresh paper. "Asphalt-impregnated felt" is the trade term. Tarpaper. Oily and tarry to the touch, good to smell and handle when fresh—you know it's impermeable and will do the job. But it tears easily, doesn't last. The site was surrounded by scraps of raunchy roofing paper. Damp inside, it became nesting-grounds for huge, striped centipedes; my flesh crawled a little when I picked up junk to load it on the truck. What lurked inside? I didn't want to find out, I just picked up bunches and threw them on the truck, fast as I could.

We finished up the metal patching that Saturday morning; cut scraps with tin-snips, laid them over bad board joints like tiny cant-strips, and nailed them down. Over this section of the deck we laid 30-pound paper, in the hope that the decking would not cut through this heavier paper. Tarpaper comes in different weights: #15, the lightest, 300 square feet to a roll; #30, with a green micaceous surface, glittery, less square footage; #90, heavy as lead, extra thick and crusted allover with red or brown grit, can be used as a final covering. It's called "mineral-surfaced felt" and contains barely 100 square feet, if that.

We rolled out overlapping parallels of paper, nailing every four to six inches along the lap. The specs say there should be a six-inch overlap between the runs of paper. To protect ceilings, as soon as finished we papered them over; but two weeks later that papering was so shabby it had to be covered, again. The nails were gray galvanized tacks 3/4" and 5/8" long, with big flat heads. The trick was to hammer them and not your thumb—not much nail to grip when you place it for the hammer-blow.

I was tacking along once when I heard a sudden, loud, whispered roar: Tom had hit his thumb and torn it open. I doctored it from the box of first aid stuff I kept in the truck. Diego Olivas caught a shovel across his nose once, but aside from that no man was really hurt in all the two years' work. I paid hundreds in premiums for workmen's compensation and liability insurance.

As we worked on the rooftops, Tom's hat kept blowing away into the arroyo. "Egh." Either it was day-end, or starting to rain, but we were always in a tear to tack down dry sheets. It goes fast. I've gained accuracy with hammer and axe by learning to fasten my eyes exactly and unwaveringly on the mark. If you can do that, your hand will follow your eye and you almost never miss. Keep your eye on the mark and your axe will split the mark. Make the mark small as a pinpoint and focus on it.

We cut tarpaper with our pocketknives, and it was hard to do. Roofers have special cutters; they seem just to stroke the knife across the paper and it falls away clean-cut.

Buford came out to inspect. His visits were growing rarer. He was impressed by the canales. "That's really a well-made unit," he said.

Electric wiring was to run overhead through the roofing. A couple of weeks before, I'd gotten estimates for the electrical contract, and Rayford Sims, of Ancient City Electric Co., got the job. I was to buzz him two days before the roofers came, but I forgot; Monday morning I had to race over to the Munros' to telephone. Rayford's brother Mack and another man got there just as the pumice was covering the overhead drops in the kitchen, and they had to work through it. They hustled and managed to keep ahead.

Farwell's men came in a convoy: two dump trucks loaded with pumice and two or three other trucks, one carrying a conveyor, another pulling the asphalt cooker. They set the conveyor, the bucket-loader, in front of our bedroom window, then backed the loads of pumice up to it. With the dumper upended, a man stood in the load of pumice and with a shovel worked and fed the light gray grits through a little door in the tailgate onto the conveyor. The conveyor bore the pumice up the track and spilled it onto the roof.

Two men with light wheelbarrows moved the pumice from there to the farther reaches of the roof, where three others were spreading, grading and sloping. Corners were shaped to run off water, and cants were molded into the angles of wall and roof. The pumice was shaped and graded in such a manner that nowhere would water be trapped, nowhere would water be allowed to stand. Generally, they built and angled corners at the downward end of a run. The upside of frames, skylight casings, et cetera, were sloped or canted so as not to trap water. Starting at canales and grading upward from them, they used a straightedge and level to check the pitch.

They worked fast. I think of them as a circus troupe, a team of tumblers or acrobats. They were wonderful to watch. Six all in motion at once, so easy and sure, no rush, no clutch or sweat. Each knew what to do, each served the others. They were men with pant legs and shoes caked with tar. One was very tall, thin as a stick and dark. When he smiled he showed an array of huge white buck-teeth. He took a roll of paper, tossed it, keeping one end so it spilled and unrolled the length of the roof. It looked so easy and deft. Another man, the foreman perhaps, was an Anglo about 40, who had the motion and graceful easy stance of an athlete. He was bareheaded, and he directed operations in a courtly, light and casual manner while he worked at one task or another. Nailing down dry-sheet, he gave the work a certain beat, feeding nails into place with his left hand, following with a tap and bang of the hammer. He went fast and perfectly sure: tap-bang, tap-bang, tap-bang, right along a line. It was something like a shoeshiner making rhythm with claps of a rag.

Another man wore a white cap whose broad visor shadowed his face and round dark glasses. He wore a coverall and big, black-crusted shoes, and he moved with a rolling motion, tall and bowlegged. He and the nail-drummer were headmen, it seemed; but all were experts. It is hard to remember the work in sequence. I picture it as going on all at once: conveyor spinning out pumice, men shifting, wheeling, grading, unrolling, mopping. I think they covered the pumice as they went, section by section, since pumice must be kept dry and out of rain. Layer on layer of 15-pound paper, roll after roll was slung and went ribboning out across the roof. White-cap dipped his mop, lifted it smoking and dripping hot asphalt, and swept mop and asphalt over the lap between runs of paper, sealing them together.

Meanwhile, down below, the burner had been fired and was thundering. A man chopped open round cardboard cases of asphalt like giant Tootsie-Rolls and piled them in the cooker. The asphalt started to melt and sink. Men carried big smoking kettles up a ladder to supply the moppers. Paper and heavy slops of hot stuff were led down into the troughs of canales, far into them. The roof must be bound into the canales, it must flow into them. The flashing collars were slipped in under the final layer of paper; jacks and casings were bound in place.

The pumice had to be laid on and sealed up in one day. So by the end of that day, the troupe had pumice plus a couple of layers of the 4-ply roof set, sealed and ready for rain. They may have placed some of the final 90-pound wall-flashing, too. (Next morning they started in again, but were driven off by rain and didn't make it back till later in the week.) It was a busy day all around; I don't know what Tom and Bonny were doing, I was so busy. The kitchen cabinet man came to take measurements, and I had to rush downtown to get another roof-jack because of an on-the-spot decision to install a vent and fan over the children's toilet. None was planned: Mack Sims advised it, and a good thing he did.

One of the final steps was the laying of the 90-pound gritty felt through the firewall, down the inner wall, over the cant and onto the roof. This was done in overlapping strips. It was sealed and secured with hot asphalt. They put this #90 around all box-casings also. They didn't try to stretch paper around corners and projections, but always cut flaps and folded the flaps over the corner in a particular way and sealed the seams with black mastic. The graveling followed as the final step. Loads of pea gravel were hauled up by conveyor. Buckets of hot asphalt were mopped overall and gravel, a thin layer, was shovel-scattered on top. It stuck fast to the asphalt. It was a thin layer but adequate to prevent hail damage.

When all was over, and the last vehicle of the Troupe had rolled away (to cheers and whistles), I climbed the ladder and lo and behold! our whole suffering project had acquired a wonderful thick, crusty, impermeable skin, all folded, lapped, seamed, caulked, tarred and graveled, graded and sloped to the spouts. It was glorious to see. It was so encouraging to have this extraordinary team of professionals come dancing in and do us up all shipshape. Even though there was lots more to do, the roofing gave me a second wind. And I didn't have to do a thing but stand back and watch it happen.

The next push included finishing the parapets, installing metal flashing, and chicken-wiring the walls for plastering.

Flashing came in 100-foot rolls of medium-gage sheet metal, 14" to 16" wide. The flashing was to shield the #90 through-wall paper, where it penetrates the firewall and where it runs up over the cant strip. The wider your metal, the more you shield, the better. It is most important to bend it so that all water is deflected and runs off. If not bent properly, if a hump is left anywhere and the water can't escape, the 'moist' will stay and work its way under plaster and inside the parapet. In many places we left shoulders in the metal, the plaster has separated from the metal, water has penetrated and the adobe inside is crumbling. The sealing and shielding of parapets is very tricky. Things we should have done according to plans but didn't: painted damp-proofing on top of firewalls, under the plaster;

told roofers to run #90 through the wall and down the outside wall; hooked metal flashing into the wall. Short of rebuilding parapets, all I can do now is try to plug cracks between plaster and metal.

The sheet metal is unrolled along the firewall and fastened down by layers of adobe, which builds it into the wall. At approaches to canales and at wall-corners, the metal has to be cut and pieced to cover the corners. The height of the parapet varied—the more courses, the more work. We were too high for the scaffold now, adobes and mud had to be hoisted by rope and pulley. It was slow. Sam Pratt had left for college, and Baca brought helpers whom he found, and I hired. They didn't last. There were three in as many weeks, one after the other, and the last melted away owing me money, after I made the mistake of advancing him some against his wages. When Weatherford asked for an advance, I knew he drank it up. But I could always get it back from him in work.

Baca had pulley, rope and hook, and he rigged up a hoist on the roof. The parapet-work and the chicken-wiring went on until October. They started the runs of "poultry netting" at the top, inside the firewall, at flashing level; carried them up and over the firewall, and down the outside wall to four inches below finish grade (ground level). They fastened the wire to the wall with four 16d or 20d nails per square foot. A lot of chicken-wire, a lot of nails. The three-foot-wide mesh was applied vertically and overlapping. At the juncture of plaster and wooden frames, where two different materials met, Francisco had me cut strips of metal lath six inches wide to cover the joints. The heads and jambs were all bordered by strips of lath, and we should have nailed them diagonally across corners, too. But didn't.

Plastering hinged on certain carpentry. Tom framed and built a sloped shed-roof over the cellar steps: 2 X 4 joists decked with plywood, covered with #90 roofing, later with shingles. Temporary bulkheads of wood frame, wallboard and plaster had to be put up to close off the hall, the dining room doorway, and dining room window openings. Windows for those openings were to be specially made, a problem I was putting off to the future.

Door and window-frames and sills had to be made and set. This work was delayed because Tom couldn't get hold of his brother's table saw and joiner, which he needed to cut sills and finish frames for exterior doors. The rough frames were in "bucks" and "jambs" made out of 2 X 6 common and nailed to the built-in anchor blocks with 16d smooth box nails.

("Smooth box" versus "common" nails: smooth box were brighter, thinner, better to work with, while the common were heavier, coarser, cheaper, all right for chicken-wiring. All nails, especially foreign-made ones, are flimsy wire-made things and tricky to use. They bend at the least resistance, and the heads tear off when you try to pull them out. Oh, for a good strong nail. "C, F & I" or "Colorado Fuel & Iron" seemed to be the best, and I always tried to buy them.)

Tom usually assembled door-heads and jambs and installed them in one piece. The finish frame, which was of clear stock and went over the rough frame, had to be set plumb and square, and every one needed shimming with bits of shingles wedged in to force it true.

Baca burned to start plastering. He would do it all by himself. Not until much later, in December, did he get, or even claim to want, any help. He was going to begin at the southeast corner and move along the east wall, around the bedroom windows. This would avoid the door-frame problem for a while, but not the problem of the sills; he needed the sills to plaster up to. Anyway, he began. He had been plastering and grumbling for a couple of days when the machines finally arrived.

Weatherford's joiner was a power-operated plane, a planing

mill, a steel table with an adjustable blade that spun and trimmed wood like a plane. It straightened edges. The table saw was a table with extensions and guides, a blade that rose up in a slit in the middle of the table, and a motor to run it. The speed and control afforded by the table saw made it possible to rip lumber lengthwise with accuracy. Among other things, the blade could be tilted to cut at angles. Once a piece was sawn, it could then be run through the joiner to make the edges smooth and true. Among the various blades for the table saw was the "dado head" that could cut a dado or groove in the face of a board.

All these things were helpful in making the subsills of the windows. Our Andersen casements, ordered weeks before, were installed and nailed to their rough frames by now. What lacked was the 2 X 6 subsill, an extension of the sill already on the casement. Designed to fit under the windowsill, it sloped down at the same angle, so the edges had to be ripped at an angle. Also, a groove had to be cut the length of the piece out near the front edge, to act as a drip barrier underneath. Water doesn't stop at the edge of something and drop, it tends to roll on around the edge and underneath; so a barrier must be made, a cut, a "kerf." Tom pushed through this installation in pretty short order, though nothing ever went quick enough for Francisco Baca.

On he came, the demon plasterer. I see him going to it, his right hand brandishing his rectangular trowel, his left hand gripping the flat square tray with a handle called "hawk." It was a hod for carrying plaster, which he swept up off the mortarboard with the trowel. He took plaster off his heavy little hod with his trowel. He tossed the hawk, tilting it toward himself, and with a backhand turn of the trowel he scraped off some plaster to spread on the wall. He tossed the hawk to keep the plaster from spilling. It was a quick motion, all in one. Skup, it sounded. Wrist-action. Twist. "Skup." His trowel-motions were all away from himself; plaster must be scooped onto the face of the trowel and kept from spilling off either hawk or trowel. He juggled the plaster on the hawk, tossed it a little back, and then caught up plaster backhand on the trowel. "Skup." And in the same motion, the follow-through, he pushed the plaster up the wall with the trowel. He was a tall man with a long reach, and he thrust the plaster upward and pressed it strongly against the wall and chicken-wire, up to the full stretch of his arm. All day long.

Plastering had to be carried from below ground level up over the parapet and down the inside to the flashing, all of which required two stages and scaffolding. Talk about baseball pitchers: a plasterer must develop arm and elbow and shoulder troubles beyond anything famous.

The base coat, or "brown coat," was a mix of Portland cement, Rich mortar or pure lime, and sand. Cement and Rich were mixed dry, then mixed with water and sand, ratio of 1 to 3½ sand, in the mixer. If hydrated lime was used, "masonary lime" as Francisco called it, the mix is 7 Portland to 3 lime. The whole house carried this rough gray hide of "brown coat" until April of 1966.

Baca let the plaster set a little, then went back over it with trowel and darby to smooth it and even it. The darby was a long flat metal straightedge with two handles. He had a large paintbrush, which he hung on his water-pail by a nail; with it he whisked water on the plaster as it was setting to moisten it again for troweling. Dirty water-drops flew over everything. Inside, muddy white drops all over ceilings and woodwork. Terrible.

Once plaster was brought in around the windows, the trim could be put up to cover the joint between window and plaster. Tom used the plain 1 X 4 that I'd already primed with white primer, and mitred the corners. When these were nailed up, the crack between board and plaster had to be caulked to keep out damp. Here I got into the puzzle of caulk and caulk gun, the metal contrivance to hold and shoot the tube of caulk. It was the devil to control once the trigger was squeezed; too much caulk would glob out, more than wanted, so I found myself trying to catch it and rub it in with my finger, first one finger, then another, and another, getting all gummed up in the process.

October stayed clear, but colder and colder. Outside wood needed to be protected while the days were still warm enough for painting. Pressed for time, I hired Gus Bokum, a painting contractor, to finish outside work while I went inside to prime interior wood.

Bokum was about 40, shaped like a barrel, with a fat neck and a bristly crew haircut. He had a confidential manner and a voice like a file. "I hope you're satisfied?" he rasped sweetly, smiling, nodding and bowing out the door. He did pretty good work, not as careful as mine, but good, and efficient. The next time I saw him, a year later, he was recovering from prostate surgery and he could hardly walk. He sometimes worked himself when he ran short of money, to save on payroll and cut down his weight. He told me how he would surprise his men by the speed of his work. He liked to tell stories that made him look good. He would get a shrewd cunning look on his face. He used his hands, eyes and head in talking. He squinted one eye and tilted his head; he shrugged, half-smiled, nodded as he talked, trailing off in a wheezy chuckle. He liked his Scotch.

Paintwork: caulk, prime, plug, sand, clean, finish, step by step. Well I know those casements, every line of molding and sash, and especially around glass, inching the brush along the muntins just touching or all but touching the glass. Once when trying to keep the paint off the glass, I deliberately punched the brush onto the glass. I or my devil did. Imp goaded me. My body braced and contorted inside the casement, my brush focusing and creeping around little muntins, I began to giggle and couldn't stop. I snickered and snorted—anyone would have thought I'd gone mad. Which I had, for the moment, and small wonder.

In my back pocket were a rag, a brush for sweeping, a scrap

of sandpaper and a putty knife. I whipped out putty knife and rag, wrapped rag around blade, and wiped off paint spatter. At this date I wonder if the house ever would have gotten built, without my exerting all the stubbornness, self-control, self-taxing severity that were in my nature, every day, week and month.

Was it just pressure of duty and will? Was it marriage and family? Or was it something else that held me inside that casement aiming myself so intently on a bead of paint. Perhaps it was the work, the desire of building, a rhythm, an effort that would result in something substantial and good, in spite of whatever troubles beset me outside it. The work embraced me.

Work kept Tom Weatherford together. *This* work, *this* building. He even said he couldn't find a job as good as this anywhere else. "Bueno patron," I heard him mutter to himself one time when he was hung-over. I suspect the work not only held him, but at times inspired him, as it inspired me. He would never admit it. Still, this work steadied him and pulled him along. It embraced him, saved him. I never saw anyone steadier than he on the job; he seemed wrapped up in it, even though he was building someone else's house, not his own.

"Do you ever beat the shit out of your wife?" he said once, with a joking look on his face. I was so amazed I just looked at him. He backed away at once: "No, you wouldn't," he muttered. He was embarrassed. I wondered: did this mean he beat his wife? Was he capable of that? I guess he was, as much as anyone is, beside himself with drink and bitterness. Maybe he was half-crazy half the time. All those children in the care of drunken parents. I never witnessed it. I never entered their world. But I sense now he was reaching out to me for something—kinship, forgiveness—perhaps feeling such remorse that he needed to touch me with it, or make me party to whatever had stained him. And I, his young boss, out of a world he could not reach, I rebuffed him. I didn't know what to say, but I didn't like this flash of self; it made me uneasy. He saw instantly he'd made a mistake with me, and shook his head to dismiss it.

In order to protect the glass and speed the painting, I stuck masking tape, miles of it, around the edges. It was a bad mistake. When torn off, it left a gummy residue that was much harder to clean off than paint. Never use it!

In Amy's future room was the odorous clutter of my paint shop: brushes, rags, turpentine, sticks, spackle, putty, jugs of mineral spirits for thinning, Minwax stain, Woodlife, linseed oil, caulk, varnish to seal knots. I used gallons of Pittsburgh white primer-sealer on all the sills, windows and doorframes, baseboard and window trim. For the finish coat: McMurtry "One Coat Flat" white, into which was mixed, on Bokum's prescription, a bit of umber or raw sienna, for warmth.

Meantime, a lot of work was being pushed forward inside the house to prepare for plastering. Electricians and carpenter had to keep ahead of Francisco, who swept all before him. Tom's task was to set the baseboard and the door casings. On the table saw he ripped how many hundreds of feet of 1 by 4 on a bevel to catch the plaster. I slapped white primer on all of it once. Then he fitted, cut, mitred, stepped and fastened segments of it together with corrugated metal fasteners; and nailed it to the nailing blocks preset in the wall at baseboard height. It was both a flooring and plastering guide. At doorways, it butted in flush against the doorframes, which he made and installed in one piece.

For the finish doorframes (casings), I bought all clear pine, expensive but fine material, with no bleeding knots. Knots seep through paint, also tend to separate and rise up from the surface. Clear stock is graded in sizes all its own: "five-quarter," "six-quarter," "eight-quarter," meaning roughly, inch and an eighth, inch and a half, inch and three-quarters, in thickness. Just an odd way of putting it, like so many things in this business. We used the inch and a half, "six-quarter," for casings. The height of each was determined by finish floor elevation and door-size, the depth was determined by wall-thickness plus plaster-thickness, on each side. That generally came to 5½"-plus baseboards, seven inches overall. Meaning I had to get "six-quarter by 8," and lose some, about half an inch off the edge. (I saved all that stripping, it seemed such nice useful wood.) Once the casing was set in place, it became a plaster ground, a guide for the plaster that would come flush with the edge of the casing.

Tom ran the lumber through the table saw and the joiner, to size, plane and rabbet all the pieces. A rabbet is a square ledge or step cut out of the edge, to make a bed or seat for glass, for panels, or in this case for a door-stop. For such precision cutting, a table saw is essential. He cut the rabbets a half-inch deep, as wide as door-thickness, all around the edges of jambs and head of a given casing. He mortised the head to lap down snugly onto the jambs, nailed head to jambs and braced the finished work until it was installed. I primed them all white before they were put up. He nailed each casing into its opening with 16d finish-nails, and punched the small nailheads below the surface with a nailset. You could see nailheads glinting inside tiny tunnels in the wood, before the holes were filled with spackle, putty or plastic wood, sanded and painted over.

The doorframes had a peculiar, rootless, midair look before the flooring came up under them. After the plastering, the 1 X 4 trim would be nailed around the frames to cover cracks and to dress up the doorways. A crack always opens up between plaster and wood, when wood sucks in moist from plaster and plaster shrinks. For trim we used our 1 X 4 flat around head and jambs, offset 3/8" from the edge. No mitred corners—mitres open up in time because end-grain is so open and porous. To my eye, the stepped edges of the doorways are very distinct and pleasing, a series of offsets—edge of trim, edge of jamb, edge of rabbet, running sharply and evenly up one jamb, across the head and down the other jamb.

Other woodwork had to be in place for the plastering. I used Kitty's drawing for inside window trim, a two-inch-wide molded strip (curved, in section), called "Streamline Casing" at Empire Builders, where I found it. Tom built the interior windowsills, or "stools" as he called them, out of 2 X 12 planks that he joined edgewise with splines, which were strips of thin plywood dadoed into the edge of each plank, like a tongue in a groove. Though I primed them and assiduously varnished all knots, they should have been insulated against the heat of the convectors underneath. They tend to cup, and the knots bleed and protrude in defiance of all varnish. Nevertheless, these deep generous sills could become seats and stages, shelves for toys and plants, platforms for everyone and everything. And so they have.

By the time of this carpentry, all the electrical wiring had to be in place, channeled into adobe walls, bored up through floors and wall plates between studs to the various outlet cans, which were set to stick out far enough to be flush with the plane of the finished wall. The runs of white cable ran along dirt, to be covered by sand and brick, and under wood floors, stapled to the joists. The cans for the exterior spotlights, one to sweep the east wall of the house, the other high on the northwest corner to sweep the north wall, had to be set in with the exterior plastering.

Our man from Ancient City Electric Co. was Wade Hoppin. Lean and slight, he walked carefully, softly on the balls of his feet, almost tiptoe. He had a lean, rather fine face with a large chin, and deep-set blue eyes with long lashes. He seemed carefully put together in form-fitting, perfectly faded jeans and blue denim jacket. He wore a blue plastic hard-hat, the only man except Mike Lujan ever to wear such a thing on the job. Slung onto a second belt low on his hip was a holster loaded with tools: spool of black tape that hung on a little chain, screwdrivers, wire-cutters, folding ruler, level, and various grips. Michael admired all the gear that Wade packed on his hip.

Much of the work he did himself, silently, efficiently. Sometimes he brought another man named Rick, younger than he and dour, with black sideburns and a black Stetson. Rick brought a radio and listened to plaintive country music: "I didn' kno-o-ow God made honkeee-tonk angels ..."

Both talked with a New Mexico rangeland twang, both had a certain cowboy air about them. Wade usually kept his clothes clean in a clean line of work, but he did have to squirm around under the floor running wire. "This is the best-built house I've seen in a long time," he told me after his survey of the underpinnings. Another time he said, "It really trimmed out nice." Always respectful, he addressed us as "Ma'am" and "Mister Gates."

There was no power in the house until December. We had to run 150 and more feet of extension cords all the way from the meter pole. Ultimately Mack Sims came out with a special machine digger to cut a slit trench from the meter pole to the house. The digger scooped the trench, while a winch and cable, tied to a juniper tree, reeled it along.

Meanwhile, George Roy notified me that the pole was on his land; we had to move it six feet to the west. Roy claimed he didn't mind, but "somebody else" might mind, some other owner. Balls! Terrible toil for a six-foot move. Bonifacio and I chopped frozen December dirt, chip by chip, with picks. It was freezing weather, but we sweated. It was tricky, uncovering the line without cutting it with the shovel. We had to break through frozen gravel residue from concrete-mixing, slabs and cakes of cement. The result of all this misery is that two lines run from the meter pole, one to the well, the other, with the telephone line, following the curve of the road then straight to the circuit-breaker panel in the garage. Both are about 18 inches underground.

On came Francisco, with trowel and whisk and a river of mud. No chicken-wire was used to cover the interior adobe walls. Adobe base-coat plaster was simply dirt, coarse-screened, leavened with sand and bits of straw, and hoed and slushed in the mortar box. For plaster over wood frame, metal lath was stretched tight over the studs in a reverberant, drumming web, with pieces cut out around the electrical outlets. The lath was nailed with "blue nails," sharp thin tacks with big heads, colored a gunmetal blue like those dark-blue wasps you see walking the ground, winking and snapping their wings.

First on the lath goes a coat of "extra-fibered" plaster, which comes full of hair, deer hair according to Francisco. After this coat began to set up, Francisco took his small scratch-rake and scored the surface, to provide tooth for the next coat. Once dry, the scratch coat was very stiff and hard. In bedrooms we put mud over the scratch coat, but in bathrooms and kitchen we went the whole hard-plaster route: scratch coat, then brown coat of regular plaster, then Keene's cement to finish. Keene's was white, and looked like a pile of whipped cream in the wheelbarrow. I remember scrap wood roaring in the new fireplace, and Francisco and Eloy Padilla talking and laughing as they troweled the Keene's whip.

For adobe finish coat, adobe dirt was screened through a fine mesh. Tom Weatherford made a sifter, and Bonny shook it back and forth, patiently sifting dirt into the wheelbarrow. When he had raised a pile, he added sand and water, and mixed and slushed. It made luscious fine mud, and a barrowful went a long way because it was spread on so thin, just a bit thicker than paint, just something to even and mellow the surface. But it had to wait for the first coat to dry.

That first mud was laid on so heavy, an inch or more in some places, that it took a long time to cure in the cold house. Mud needs to cure evenly—too rapid in drying and it crackles. And if it freezes, it just falls off the wall. I rented a kerosene heater to push along the drying; I was worried the mud would freeze. With no heat in the house, we had to open the valve under the bedroom floor every night before leaving, to drain the water lines.

I recall coming out one night to change the heater. Inside, my flashlight probed the hall of the dark mud house, flashing across taped, spattered windows and along trashy dirt floors where pipes lay still exposed without their insulation, in their trenches. The place was clammy, cold, smelly, awful. But outside, how dark, how still it was. The stars! It was deep country night, no streetlights or traffic, black dark except for a thousand piercing bright stars. It was one of the wonders of my life. So quiet. One lone light far away across the arroyo. This was where we were going to live for the rest of our lives.

Plastering dragged on and on. Baca wearied of it. It drove him crazy coiling himself up to plaster closets and small, cramped spaces. Interior work was much more complicated and tedious than exterior. The dining room, his last room of the year, shows he was running out of steam; it's crude compared with the rest. Was he trying to stretch the job by doing it all himself?

He finally agreed he needed assistance. Eloy Padilla, the man I hired to lay the brick floors, helped him out in the kitchen. Eloy: glossy black hair, long and curly in back; splayed teeth; thinly dressed even in the cold, but wearing a fur "trooper cap," with folded-up visor and earlaps that folded up over the top. He was a sure, easy workman, with lots of palaver in a low easy-running voice that wove its way into your confidence.

"Pretty soon we go home see Mama," Francisco said in comic style when rain threatened.

"Go home and play with Mama," as Eloy put it another time. He seemed lecherous; his laugh was sinuous, "Heh, heh, heh." He talked of "neegers," and he liked to repeat his punch lines. "Maybe 'Mister' Runyan will use the 'master' bet-room. Heh, heh, heh." The idea of 'Mister' Rowena Runyon tickled him. He had worked for Rowena Runyon and Grace Gibbs, and he got a lot of mileage out of that situation.

"Breeks" he said. Six thousand bricks were used in the first half of construction. The Kinneys were the cheapest, weakest bricks; the reason Ann and I wanted them was that we liked their color and texture the best of all we saw. They were a nice soft pink, muted and not harsh like most others. They cost 6¢ apiece, and were all delivered from Albuquerque at once by a huge truck and piled outside the bedroom window. Eloy's fat silent son Fred helped him lug them in and work them into a floor.

Ann composed a pattern for them, a basketwork of pairs of bricks butting perpendicular to other pairs, ends to sides and sides to ends. The basket-weave was crossed at intervals of four bricks by a band of single bricks end to end, then more basketwork and another band, and so on all around the house. In most areas, Eloy carried out the pattern very well. Trouble came later because the setting-bed of sand and dirt had not been compacted properly. Sand bedding for bricks must be no more than two inches deep, and must be thoroughly soaked and tamped, as does the dirt underlying it, or it will settle and sink later. Bricks have sunk pretty badly in the kitchen and elsewhere. There was a hump near the laundry where piping was laid so shallow it buckles the brick like the root of a tree.

To level the areas of sand with each other and keep the floor in grade, Eloy used a long 2 X 4 staked up to elevation, which he called a "screed." His cutting and fitting was quite nice, witness the row at the top of the kitchen steps, and the angle across the curved front of the fireplace. We got the idea of cutting down some of the unevenness by machine; I rented a rotary grinder, and Eloy ruined some brick in the kitchen before I called a halt. It polished off the texture—no good at all.

What a difference, what a lift the floors gave to the place. Once they were established, it began to feel like a dwelling. All the wiring, piping and dirt were covered up, and block foundations disappeared. The door openings looked like real doorways. We rose closer to the ceilings that seemed to come down more cozily over our heads, and everywhere our feet felt the comfort of a firm level base.

For hardwood floors, we used random-width, tongue and groove white oak produced by Bruce in Tennessee, and distributed here by Kimball-Jenkins, who installed and finished. The boards came in random lengths in heavy banded bundles. The lumber, some of it seven inches wide, was "backed out"—grooved along the backside, perhaps to help resist cupping or cracking in expansion. In this material, we again departed from the specs, which called for the standard two-inch red oak strip. Another happy departure, another blessed avoidance of something chinchy. The stuff we got was the heaviest, hardest wood I ever saw. Ordinary nails curl up and die trying to penetrate it, and it would split and splinter unless special care were taken. Screws would split it unless holes were drilled first; it required a special nail, a flooring nail. Think of the ancient British Navy, built of oak. To saw it was like cutting rock, and it gave off a sour smell when sawn.

Jimmy Jenkins was the floor finisher. He wore a cap and khakis, and walked lightly; he was small, with small teeth like baby teeth. He wore glasses on a round baby face. He was about 45. He had folksy ways; every sentence, he addressed me by my first name. "Oky veedy, Bill," he said, "Oky veedy, boy. Right, Bill. But Bill—Bill—it'll take three weeks to get that plank, Bill. Oky veedy, boy."

He told me that good white oak flooring was hard to get, because the good white stock was usually sold to furniture makers. He said gym floors were usually maple, sometimes beech. Jimmy drove himself hard. Up early, finishing floors all day. Floor finishers made about $4 an hour. They earn it.

At my direction, he left bedroom floors unsealed. Sealer darkens the wood; we wanted natural floors, as close as we could get by simply waxing them. The flooring was laid by a lanky, short-tempered Texan, who drove up one day and said, "Somebody want some ranch plank laid?" He had a drawl and nonchalant motions, but did his work with furious impatience. Sling, crash, pound—no finesse. Another man followed several days later to screw and plug. He drilled and countersunk holes on one run, whizzed in screws on another run, pounded in plugs on the final run. It was meant to look like pegged flooring; it's a fraud and I never liked it. The next round, in '65, would be installed without "pegs." Jenkins expertly sanded and finished the floors.

We chose a pebble-pattern gray vinyl for the floor of the children's bath, and it was installed by the supplier with mastic over a 5/8-inch plywood deck that Tom put down over the subfloor.

With plaster and floors finished, Tom began setting the base-shoe and hanging the doors. "Base-shoe" was the quarter-round molding meant to follow the baseboard in all its ins and outs and miles around the house, to cover any discrepancy between base and floor. It had to be primed like all other wood.

Doors forced me to two alternatives. Stock doors (factory made) came only in heights of 6'6" or 6'8", in widths of 2', 2'4", 2'6", 2'8", or 3'; and the range of design was very narrow. But they were well made, cheap and handy. Other than that, I had to go to Werner Hildebrandt, the only real custom worker in Santa Fe or Albuquerque. He did nice work, very expensive.

Hildebrandt was a tall Nordic fellow with a square blond head, flaring nostrils, and a square jaw that widened when he grinned. His grin revealed two large canine fangs on either side, longer than his other teeth. He talked with an accent, and smiled as he talked. When he agreed, he would suck back a "Yup, yup, yup," as if he was choking on them. Unlike any other workman, he moved in Santa Fe society; one saw him at parties, one asked him to stay for a drink. I tended to visualize him under a helmet or a Wehrmacht officer's hat.

Good as his work was, we sometimes found a botch or a goof that would have to be made good. And if you provided your own design, he would take it and use it in his trade without asking you or giving you credit.

Most of our doors I got from Santa Fe Builders, stock-pattern six-panel and four-panel Curtis doors. The four-panel had to be specially built from a discontinued Curtis design. The big manufacturers keep eliminating models, narrowing production, imposing conformity. Even the other "custom craftsmen," such as the famous Ernie Knee, only made doors in a given number of patterns of their own, and would not deviate for love or money. The trend is toward bigness, speed, mass production, squeezing-out of craftsmen. Make buyers conform.

It was my idea to stain the doors. I used Minwax "Natural Oak" or "Light Oak" stain, two coats. The "Light Oak" is darker, but neither has anything to do with oak; they are just shades of brown. I controlled the stain to let the original reds and yellows of the pine come through. I did fourteen doors in the garage at Allendale Street. It was October, chilly, leaves blowing, and it was good fun to see what the stain would do to the wood. Summer wood, spring wood: see in the face of the pine the stripes of light and dark, the light being soft spring growth, the dark being hard, narrow summer growth. The spring stripe is more open wood, and staining it reverses the coloring, like a photo negative. Stain sinks into spring wood and turns it very dark, but doesn't penetrate the hard summer stripe, barely colors it at all. "Into something rich and strange." In those doors and their colored patterns of grain, I could see clouds and moods of sky, winter sunsets. Intense drinking focus of my eyes on things that I would not notice so much afterwards, like the pine doors or the bricks.

To fit and hang the doors, Tom cut them crosswise for proper height and planed the edges as necessary for width. He nailed two pieces of wood in a V at the end of his sawhorse, to hold a door while he planed it. He had a "butt marker," a metal template in the size and shape of the hinge-leaf, to mark out and impress an outline hinge-deep (about 1/8") on the door edge; then he chiseled out an exact seat for the hinge, to mate with another similar notch on the door frame. Thus the hinge-leaves would snug in flush with the wood.

Ann designed and Tom built Michael's bureau in the little space between the twin closets. It's mostly of clear stock, with plywood drawer-bottoms dadoed into the sides. It is nice wood, but too soft and easily marred.

Tom's masterpiece was the channeled trim for the skylight openings. He cut the pieces out of 2 X 2, into quarter-inch-thick right-angled channels of pine; then he fit them around the openings with mitred joints at the corners. They fit perfectly then.

"Beautiful job," I said.

"Yeah, they came out pretty good," he said. He was as proud of that work as anything he did.

He followed Kitty's drawing in constructing the wooden boxes for the overhead lights in the bedrooms and kitchen and hall. Where necessary, he scribed the lumber to fit the rounds of vigas. For a scriber, he used a compass and pencil, drawing the compass point along the uneven contour while the pencil marked a similar line on the lumber. We didn't like Kitty's idea for radiator covers, so Ann designed a simple grilled cover, which Tom made like a door, cutting and piecing together vertical and horizontal members, then gluing and tacking in the little bars. Another difficult, tedious job that he coped with. He shelved the linen closet and laid up shelves in all the clothes closets, supporting them with 1 X 4 nailed around three walls.

To go back and pick up the intestinal matters of the house:

A sewer system had to be provided, instructions for which I called the County Health Department. That fall, two agents came out and showed me how to take soil percolation tests by digging several 30-inch holes, filling them with water overnight, then filling them again next day when the agent returned, so he could measure the drop-rate of the water. From this the agent determined the percolation capability of our earth and the system we needed, be it drain-field or seepage pit. Down below the prevailing caliche in our area, we ran into deep sand and gravel, which was excellent for percolation; so we were directed to dig a seepage pit. One time a visiting Vietnamese nurse came with the agents. I recall her face, her silence, when told that only five people would live in this huge house.

Once again, I called in the backhoe to dig the great holes for a septic tank, for the seepage pit, and for the 500-gallon heating-oil tank. The four-inch sewer line I had been drudging at all summer finally terminated at the septic tank, a prefabricated concrete box made by Colony Materials and delivered by a truck with a crane. The septic tank was to overflow into the seepage pit, which was dug about 20 feet east of it. Per the County instructions, I went 15 feet deep with the pit, and built in it a structure 12 feet in diameter, with dry block walls, open-jointed, igloo-style, which my men topped off with a poured, reinforced concrete cover.

Within the rough enclosure formed by tank and pit, and closer to the wall of the house, the oil tank was buried. Two 3/4" pipelines were run from the oil tank around the projected south and west walls to the basement and the furnace: supply and return lines necessary for recirculation of fuel. Now about six inches under dirt, paralleling the outside of the house; then in late '64, they hung in midair, draped across trees by Sena's men, to bridge the deep gully that used to lie under the present living room.

Willie Sena, "The Plumber with a Conscience," was written on the side of his truck. His voice was low, plausible, peaceable, his stories were full of hard luck. Only once did he lose his grip, when I refused to pay him the final sum.

"Well, you get Mr. Bartell to inspect it and when he's satisfied maybe you can pay me—and I hope you have a happy Easter," he snapped over the phone, and hung up. He was in the cooker, under pressure, and it made him evasive. I felt he was making a sucker of me. He was always sharking for money. I gave him money until most was paid up, yet the job was far from complete, far from right.

"Well, Bill, I'm roughed in …" he said, prodding for the money due after the first phase of the plumbing was done. One time he said his payroll would bounce unless he got a payment. The last time I told him to go get it from another job, I would not give any more until Bartell had inspected. That touched him off: "Hope you have a happy Easter," et cetera.

He was newly in business for himself, and he took on more work than he could handle. He got in over his head. He didn't know oil furnaces; he didn't know how to run his business. Question: did he only half read or comprehend the specs I gave him to study so he could make up his bid? Or did he figure he could get away with short cuts? He said he lost money on our job. I thought he was trying to trick me; the evidence seemed overwhelming. But I can't make up my mind about him even now.

He wrote Ann and me a letter, an apology for delays: stick with me and I'll take care of you, you've born with me through my trials and I'll make it up to you. Honest. And I was the only one who didn't press him hard, who showed patience. He was being penalized daily for delaying his Armory job. One day I got a phone call from Santa Fe Builders' business manager: Willie owed them big money; had I paid him off yet? If not, I could just make out his check to them, or else they would put a lien on the job. That burned me up. But when I told Willie, he said it would be all right with him.

He was a big, fine-looking man with a frank, honest appearance. He looked something like a fleshy Charlton Heston, without the aquiline nose. "To be honest with you, Bill ... " and "But ... if that's the way they want it done, well, then that's the way we gotta do it." That kind of remark was common, delivered in a slow, pausing, thoughtful manner. He would pause before answering you, collecting himself. "Oh, I'm not trying to talk you out of it, Bill. I'm not trying to get out of doing it."

Our job cost him because Bartell and I caught him up. He had to tear out whole runs of soft copper tubing; the specs called for rigid copper tube.

"Well, I goofed on that," he admitted. After that episode, I read the specs carefully and tried to hold him to them. He knew I was no professional. And in my ignorance I was tempted to accept his word.

"What hydrant, Bill? What hangers, Bill?" What was I referring to, what could I mean? I had to tell him over and over to do things.

He claimed his suppliers gave him trouble. "Maybe you better talk to them, Bill, I bet they're giving me a runaround."

That was probably true. It was a foggy area—who said what, who ordered what and when, whose responsibility it was. Everybody had an alibi—Brown Pipe Supply where Willie did his shopping, the supplier of the furnace, the engineer who designed the layout, and so on.

It's a wonder anything works, and it doesn't work very well. Air is trapped in the lines, the convectors have to be bled frequently. The site we chose is windy, the house is drafty, and the convectors, tucked behind covers under windows, do not heat the rooms efficiently. I have left a certain little hex-wrench permanently in position to bleed air out of the fuel pump at the furnace intake; so many times has it failed to start because of air blockage. We should have built a damper into the fireplace. But even with our omissions taken into account, the experts have a lot to answer for.

Wade Hoppin fixed it so we could hot-wire the system at the thermostat outlets; thermostats could wait until the walls were finished. At last there was heat. We didn't have to drain the pipes any more. (Lately I've measured the supply line coming in from the well: it is one-inch pipe instead of 1¼-inch, as the plans specify, which might cause some of the pressure and volume-loss we experience. If I'd only checked at the time!)

Nonetheless, we moved in and for some time lived with the ache of unfinished plumbing work. I finally jumped on Sena and told him we had to have bathtubs operating for health and medical reasons. He came out himself after dark and finished them. He was working alone at that stage. When he could afford it, he hired one man, Dick Marquez, a cocky young gum-chewer, an okay plumber but even more stubbornly evasive than Willie. He never owned up to it, but I know he scarred the blue Formica countertop of the children's bath when he set in the wash-basin.

Marquez installed the heating lines. He used a pipe-cutter and threader, a machine on three legs, to cut the steel piping and to thread the unions, tees and "Monflow" couplings. Joints and couplings, or "unions," in copper hot- and cold-water piping were soldered, or "sweated." To join lengths of cast iron sewer pipe, molten lead was poured from a ladle all around the hub, over the oakum caulk.

Marquez' work held water. When the lines were finally pressure-tested, we found no leak in the heating lines. Buford and I went down under the floor through the trapdoor in the closet of Sarah's room. Belly-down, inch-worming and grunting around in the dirt of the crawlspace, flashlighting every coupling, we really gave the system a going-over. We ran a finger around every joint and union, but not a drop could we find.

In spite of that, Willie's pressure jack was showing a drop in pressure. There must have been a leak in his jack, because we checked and double-checked. With this jack, Willie pumped air into blocked piping to test for leaks. I made him hold a pressure of 160 pounds. He claimed the customary test was at 125 pounds for five minutes. It was Bartell who demanded the 160 pounds, though he conceded that sewer pipe would never undergo such pressure in normal use. But he said if water could leak, so could sewer gas. Jacking up the pressure was heavy work for Sena. He puffed and strained, then stood up with a loose, exhausted face. "I'm getting too old for this, Bill."

We tested the heating lines, then the hot and cold water, then the sewers. Sena blocked the sewer at its low point and filled the piping with water through a vent in the roof.

"Looks like a leak every damn joint," Bartell peeved, as he walked around checking the various sewer couplings deep under the floor. Wordlessly, patiently, calm-seeming, Willie followed him around, tamping and packing the lead tighter.

At the top of this system, "the top of the line" so to speak, were the toilets, or "commodes" in trade-talk. They were emplaced upon stubs of four-inch, cast-iron soil pipe that rose through the floor. There is "rough-in" plumbing and there is "finish plumbing."

But what you sit on is really a four-inch pipe, and all the rest is flush and propaganda, a mellifluous swirling and gurgling, a delicate cough in its throat when it empties. Toilets are designed to charm the eye as well as the ear, and come in pastel hues and elegant ovals. But it all goes down that fat pipe, and that is what you perch on—like the Latin poet's description of woman, a temple built on a sewer. "Oxford," "Placid"—a new model every year. Washbasins are "lavatories" and come in models named "Countess," "April," "Crown."

All this, as well as doorknobs and drawer-pulls (porcelain? brass? wood?), bathroom tile, cabinets, countertops, brass or matt bronze finish for hardware—all had to be pondered and selected, all required trips to Albuquerque. About the only fun thing to come out of all those treks was *Doctor Strangelove*. And:

"IN YOUR GUTS YOU KNOW HE'S NUTS"

I saw that on the back bumper of a car in Albuquerque. It made my day. I had seen enough of "Goldwater 64" and "IN YOUR HEART YOU KNOW HE'S RIGHT." (I had a button that said "BURY GOLDWATER.")

With heat in the house, with the floors established and the plastering finished, Bokum's painters came in mid-December to finish the interior woodwork and to kalsomine the walls. All door frames, exterior doors, louvered closet doors, French doors, all windows, sills and trim, all baseboard and base-shoe, were painted with the flat white oil-based paint. Preparation took time. With handfuls of putty or spackle to keep it warm and pliable, the men went around and filled cracks and nail-pits; next followed sandpapering and cleaning of surfaces. Then the paint went on, applied with big brushes, even around thin sashwork.

Mud walls bloomed in white kalsomine. They rolled it on in swaths. The first coat looked thin and milky blue; it took two coats to cover. Then, oh then, what a change! Even at night, with lights out, in the dark I could see the seamless white walls glowing.

Tom set about installing all the Schlage door hardware we had bought in Albuquerque. For the most part we chose the matt bronze finish; and we were pleased with the elegant volute handles we found for the French doors. According to which way the doors swung—all planned in Kitty's drawings—the locksets were right-handed or left-handed. Some were "passage" sets, not locking, some were locked with keys, and some were closed with mere friction catches. These operated with spring-loaded catches set in the door edge, and didn't require all the drilling and cutting the others did. We used these on double closet doors, for instance; and in such case, one of the doors had to be anchored with barrel-bolts top and bottom. French doors, in pairs, we bolted the same way, punching holes in the brick for the bottom bolts.

As the paint dried, Wade Hoppin fixed the covers on the electrical outlets, emplaced the nightlights over kitchen and hall steps, and hung the combined heat and spot-lamps in bathrooms. He also wired and placed the special wrought-iron wall sconces that Ann designed. She made scale drawings of two different light fixtures, one round and curlicued, the other diamond-shaped and angular, which Tony Taylor of the Old Mexico Shop took to a smith in Mexico to execute. They were perfectly crafted, and with their cylindrical glass globes they delighted us both. We were also pleased with the white globes suspended in the bathrooms and the square white ceiling fixtures in the bedrooms.

(From Tony Taylor also came the Mexican tile for the kitchen counter and the huge soft pink Mexican bricks that were laid up in bancos either side of the kitchen door steps.)

The specially ordered kitchen range-hood had to be wired for the light and fan inside it. That hood is just low enough to catch me across the scalp, just high enough to clear Ann's head.

Christmas Eve I paid off Eloy and Bonifacio, who had been completing the brick floors. There remained to clean, seal and wax. Sand was scattered over them to protect them and also to fill the cracks between them and pack them tight. The first week in January I swept and brushed and stuffed cracks with a knife blade. I picked and scraped out spots of paint and plaster. Then I brushed on two coats of Thompson's Water Seal. Hand-waxed them with Trewax paste wax, down on hands and knees, wiping arcs of wax with a rag, buffing with a clean rag. Under my hand, their color deepened; and each brick showed variations of red to orange, fiery shadings like Mexican floor tile. I saw arcs like saw-tracks. They came alive under my hands. They appeared to be another treasure like the ceilings.

Cabinets were installed just before we moved in. They were made by May Millworks, before May lost its identity in a merger with Coda Roberson, a tract-housing developer. Originally, we ordered cabinets from an outfit called "Form-a-Fab," which we always called "Form-a-Flab." On a later trip to Albuquerque, I dropped by F-a-F and found the place shut tight with a warrant posted on the door, which announced that they were under lien for Federal tax evasion. I just happened to discover it! Nobody thought to tell me, of course.

So after squawking and flapping around over that, I lit upon May Millworks. I asked the manager at May for a guarantee on their work. He got huffy, he refused, he made a speech: they were not out for the quick buck, they were not fly-by-night, "... been in business since 1959." And besides, he declared, a guarantee was only as good as the man behind it. So, no guarantee. I should have been suspicious. A few months later, May merged or was swallowed up by Roberson, to do work for Roberson only. I had to stumble on this indirectly; nobody told me. In a panic I called and talked to the same man, Mr. No Guarantee, and he assured me they would do our order. The man who'd taken the original order and measured the spaces in September was "no longer with May Millworks." It's a miracle anything came out right.

At the last minute before we moved in, two men came roaring up with a truckload of cabinet parts and countertops, trim and tools and stain, and worked like mad putting them together and installing them. They kicked them and wrenched them into place. It was a crazy operation, and it gave a surgical view, too close for comfort, of slam-bang jerry-built cabinet-work. We lucked out in our choice of Formica colors, chosen from tiny samples: "Signal Red" for the kitchen countertop, "French Blue" for the children's bath, "Pumpkin" for our bath, white for the sewing room and the laundry. They worked!

Cabinet doors hardly bore inspection. Some were just veneer over "composition board," a conglomerate of sawdust, chips, offal mixed with glue. I was dismayed. Other doors were better: birch veneer over plywood, and solid birch glazed doors. The men trimmed every joint, scribing the trim to fit uneven walls. They used a miter-box to cut angles, and a saber-saw, a small power jigsaw, to zip through everything else. They rushed along, crashing, whacking, countersinking nails and filling every hole with putty stained to match the finish. In their frantic mass-housing way they were very efficient.

Meanwhile the custom-work was proceeding more to my liking. Werner Hildebrandt made twin beds for Michael's room, after an original by Alef de Ghize. The paneling of the head and footboards, the head peaked like a pediment, was excellent. Werner made the tall ¾-inch cherrywood doors that Ann designed for the niche in the angle of the hall, and supplied the cherry casing to go round them. (Next year he would build and panel the large boxes with lids for that same angle, for children's storage, and we would paint them ochre-yellow.) In the dining room, he provided a cabinet door with a round-arched panel, deep-set, which we painted gray.

Schofield Pratt made the huge chopping block and the round oak table for the kitchen and the walnut sideboard for the dining room. He not only gouged us extra money for the latter, but made it too wide, so that the wall had to be chopped out to get it in. It's a fine piece, nevertheless.

We moved into the house on January 9, 1965. One of our first nights there, Michael was sick in bed with Ann, I was in the kitchen, and as we lay we heard high yips and a high, shivering cry. Michael perked right up. The coyotes were singing us into our new place.

Tom Weatherford and his table saw moved in with us; the hall angle was his workshop. Fourteen inches of snow fell one night and buried the aspen poles outside, but he dug them out and hauled them indoors to continue work on the ceilings.

(In August I'd driven up in the mountains with Francisco Baca to gather dead aspen for latias. He showed me how to gather them on the up-side of the road, rather than the down. We heaved poles like spears down through the forest, from one collection point to another. It was hard work, threading them through the trees. We got all dead ones, full of color, no need to peel them. Dark and gloomy up there, rain always falling or ready to fall, made it greasy and slippery underfoot. We got many poles too thick, 4 and 5 inches. Tom had to rip them to fit in over vigas, and they look too thick for those skinny vigas in the hall, those "squimpy" vigas, as Olson would say.)

We decided to simplify things by laying latias straight, not zigzag; but even so the work, the screaming of the saw as it battled through the aspen, the sour burnt smell and the fog of dust, filled the house for weeks after we moved in. With that and the hangover plumbing, I felt I was stuck in an endless drudging process. The house did not seem good, it seemed ugly. The windows were filthy still, ceilings were spattered with paint and plaster. I saw no beauty, only blemishes.

Finally, construction went on hold for the rest of the winter, while each of us in varying degrees tried to make ourselves at home in this structure that was to be ours forever and ever.

III

About mid-April 1965, we began the second push of construction, to include a living room, guest apartment, hallway, front entrance, patio with portal, garage, stable and corral.

This would complete the quadrangle of our house. It was a big hunk of work, but this time through it seemed to roll along, easy and light. This time I felt in command of the job and the plans. This time, no deadbeats, no Willie Sena, no Diego Olivas, to contend with. The main men were so good I hardly had a worry. They knew me and Ann and our ways, and we got along. Tom got drunk only once that I recall. I'd finally seen my role was to oversee, not to work. So I did little of the work myself, but made sure to have all components ready when needed, and subcontractors ready to integrate their work. I understood the sequence, and the job was woven.

I started draw-knifing vigas in a thaw in February.

By then I'd marshaled various timbers and planks—logs for vigas, rough-sawn 4 X 6s for portal ceiling joists, 6 X 10s for the living room, shorter timbers for lintels, and 400 board feet of rough-sawn stuff for portal decking. Much of the time I thought wood, weathered wood. I hoped that if I left it outdoors all winter and spring and some of summer, it would gain some color. The vicinity looked like an eccentric lumberyard, much odder than Olson's. No stacks; each piece laid out, blocked up off the ground, every batch ordered and aimed toward a certain destiny. It was a picture of my mind, dreaming of wood.

To peel the logs, I blocked them up to sitting height and sat astride them on a board cushioned with a scrap of foam rubber. I found them good to trim when dry, tough and ornery when wet. These had already been rough-scraped, so it remained to clear off the dark red-purple inner bark. I learned to reach far and pull long even strokes with the drawknife, not short choppy strokes. Each stroke made a facet. Cut down the ridges between facets, make many fine facets, the more the rounder. Fresh-peeled pine, round and carved, smooth and glossy as satin and white as a candle, beautiful to press my cheek to and to smell and stroke. Profound sweetness of clean wet pine, attar of rainy pine. The best poles, the best two or three, were straight as masts, with small tight knots and long hair-cracks, fine lines drawn along them, straight or twisted in a slow spiral. They were a pleasure to peel. Others were tough, humpy, full of harsh knots that made the drawknife drag and skip and cut too deep. I peeled eight 12-foot vigas and four 14-foot vigas. They rang when I struck the butts with a stick: donk, denk, dank.

We began the groundwork in April. It proceeded as before, with batterboards, lime-lines, excavation. Footings must connect with previous footings, so the trench for the 24-inch west wall was dug right down to the base of the cellar. In line with that wall, the living room fireplace was also based down at that level. We had to delve underneath the gully that ran downhill at that point. From there, following the natural contour, the footings stepped uphill toward the front of the house.

More concrete and steel for retaining walls between wood floors and brick, and for an adobe wall between living room and guest room. There was a two-foot hike in elevation, living to hall, and a two-foot drop from hall to guest, so there must be footings for four steps each place. Footing for a one-car garage to be added to the southeast corner. (A two-car was planned, but we halved it to save a piñon that grew close by.) Footing for the big sloping arm of retaining wall that juts out next to the front entry.

Next came block-work to lift us above-ground. Along the west perimeter, the foundation was so deep we built, as in the cellar, two parallel walls with a hollow between, to be capped by a poured slab.

Since this time I was resolved to be overseer, not laborer, Tom Weatherford brought me a good man from the start, whose name was Tranquilino Romero. He was nicknamed "Pipas," because he always had a pipe in his mouth while he worked. I still see him, pipe in mouth, nodding about in his jerky, old man's duck-walk. At 65, he still had all his hair and it was black with no trace of gray. He was so cheerful, so diligent, he would take on any work, no matter how heavy. He had a piping croaky voice, and a brown face with round cheeks like a chipmunk's. He would grin brightly at you, showing his own teeth, no false. Wore overalls and an old felt hat with the brim turned up. Vigor and sweet temper seemed inexhaustible in him.

Sarah spoke her first Spanish to him. "Como te llamas, tu?" he said. "Me llamo Sarah," she answered. "Ohhh!" he crowed and laughed. Michael messed around and Tranquilino chased him away with a large board. The Bacas, who called Michael "Dennis the Menace," were watching and they cracked up.

Tranquilino's English was not so good. "Okay, boy," he said to me (he called me "boy"). "I tell to Tommy, he tell to you." Thus we worked out a deal on social security: he was boxed in by F.I.C.A. restrictions on earnings and wanted me to stop reporting his wages, which I did.

Tom's son needed the family car for a job at the opera, so Tom rode to work with Tranquilino. Pipas had an old, well-preserved dark-green pickup, with a large dark-green chest in the back. The chest contained his tools—shovel, trowels, scoop, et cetera—all painted blue to identify them as his. He lived out in Agua Fría village. We met his wife once: black-haired, spectacled, a bingo fan who traveled to games far away, and made $1,000 one year at bingo. She and Pipas came out to visit one Sunday. Odd for me to see my workman all dressed up, hatless, black hair brushed and gleaming.

"Put'm here, boy, " he would say. "Put'm up here." Him, meaning it. He wore rubber boots to wade into wet concrete, to work it and float it. He recalled horse and buggy days, and he will likely go on working until they carry him out.

My other laborer was Mike Lujan, from the San Ysidro Pueblo, who had already been doing odd jobs for me earlier that spring. He cleaned windows and he built the enormous playpen for Amy out under the trees. (She hated it and cried every time she was put in it, so we gave up on it. This was her runaway phase, and we were trying to let her play outside and contain her, both.)

Mike Lujan would brush off the least speck of dirt that got on him. Any trench he dug was clean as a whistle, and it would take him four times as long as anyone else to dig it. He fussed and dawdled. Once I asked him to help me unload a big timber from the truck. I waited and waited. He didn't come. I went back to tell him again, and Francisco, who'd seen it all from up on the scaffold, said "Here, I'll help you," and jumped down to the ground. Francisco had no use for Mike's ways. As he said, "It sure makes the day go slow."

Mike could spend an hour pecking and fussing away at nothing. He never stood completely idle; it was an elaborately busy-looking idleness. But he was not just lazy; he thought he was too good for manual labor. I finally let him go in August; told him there was no more work for him. I was eager to get shut of him. He got on the others' nerves—ate by himself, kept his car locked all day, suspecting the others, the Spanish, would steal from him. He wore a shiny metal crash helmet on the job.

No matter what I told him to do, he would look as if it were a fatal shock. He would look at me, then past me. "Aha," he would say, and look at the ground. "Mhm." He looked at the mortar-box. He made little fiddling movements, pulled his gloves tighter. Two or three times he would say "Mhm. So you want me to put it right here. Well, I guess I cou't do dat." Then he would ponder his methods, his strategy, then compress his lips with an air of resolve. "Uh-huh. You want me to stack dese adobes. Like dis. Mhm. Uh-huh." He calculated; he turned a serious gaze upon adobes, upon his gloves, upon the pallet, then back and forth from pallet to adobes, measuring distance. measuring time. Five minutes later he might make a move. His belt creaked importantly when he bent over. Oh, I had to get rid of him, he had to go! He was full of long stories about his war travels. While he was away, he said, the ladies of the pueblo stayed behind "to keep da home fire burning. Ah-heh-heh!" (laughter)

Rough-in plumbing got done early because I had the plans in hand and went straight to one person and no other—Roy Butler. I knew he'd be more expensive than Willie Sena, but I knew I could trust him. Butler was a man in his 60s, a weather-beaten, turtle-mouthed plumber who knew boilers and hot-water heating from way back. He handed me his estimate: nearly $5,000 for not half the work of the first phase. I thought it was high, but I took it. I knew his work would be good, and I could rely on him, and that was worth thousands to me.

He assigned two good men, Ray and Fred, to my job. What had to be done: guest bath, including shower, toilet, basin, and guest bar-sink, and another smaller hot-water heater, and hot and cold water lines for these; cold water supply for the corral; and a third heating zone of eight convectors, fed by another Red Jacket recirculating pump at the boiler. Not much sewer line to dig this time, about 20 feet from guest bath to the main that ran from kitchen to septic tank. It would run just about under the front doorstep. (I used to grumble later that they didn't pitch the line enough, because "la commode" was so sluggish in its flushing. As Carol Kurth said later, "It doesn't seem to want to go down, it wants to come back up.")

Francisco Baca joined us in May to lay block, so I had a crew of four—Tom, Francisco, Tranquilino and Mike. They formed, steeled, poured and graded me a rough four-inch slab for a back kitchen door patio, extending from the retaining wall to the cellar steps. Francisco would later finish it with Vermont slate that I found in Albuquerque, another material from far foreign parts that charmed me.

Around this time I got a passion for old Eastern Seaboard houses that I saw in a book. My mouth watered at exquisite moldings, panels, fluted columns, fan windows. I wondered: what if …? could I …?

I took on Sam Pratt to build the corral and fencing for the horse we were buying from E. J. Evangelos, Holly Roberts' bay gelding with white socks and blaze, named "Hondo." Tom framed up a small stable on piers, small spot footings that we poured and blocked up at about four-foot intervals. Water had to be carried or run through hose-on-hose-on-hose-plus-buckets from the house until we could arrange plumbing. When the backhoe came to dig the four-foot-deep water-line trench, Ray laid down the flexible copper tubing right away. Electrical cable was laid in with it, and the backhoe filled in the trench as soon as it finished digging. Coordination! Speed! Nothing like it. The ditch had to be completed by hand at both ends, however—a Mike Lujan special.

For stable siding I knew what I wanted, but had to forage around the county to find it. Down in the village of Rowe, under the great Mesa, I found a small open-air sawmill. There were heaps of scrap and barky slabs, all red and black and shaggy, just the way they came off the tree. I got three loads at $3 a load. No one was there when I drove up. There were some ramshackle red adobe hovels; the place seemed deserted, though the sawdust pile was fresh. I knew someone would come if I stayed a while and went to loading my truck. Finally came a big, mustached, jolly-pirate looking man who swung in off the road in a pickup.

I drove my loads of this gorgeous uncouth slab back to the house and to Tom, who had to cope with it: rip it to straight widths, cut it and fit it to his framework. It was even more fractious than rough-sawn boards. But he managed, and gave the barn a tree-hide. For roof, a simple overhang sloping south, which we shingled for looks. (Shingles are used in New Mexico for shims, not building.)

When block foundations finally rose to floor-bearing height, a concrete top was poured over the west wall, including a large section for the fireplace that jutted into the living room. Did I call for the ready-mix truck? If not, it must have taken 100 wheelbarrows, 100 mixings. For crawlspace access, we left a small square opening low in the west wall, to be closed later with a screwed-on panel. (This because of the oversight during the first phase.) Another opening connected the living and guest crawlspaces, through the foundation of that wall.

Piers were spaced and built, and another threefold girder was fixed on them the length of the living room. Wall plates were bolted down on the block walls, and 2 X 8 Douglas fir joists were strung across the spinal girder, with X-bridging to chain them together. Then 1 X 8 subfloor planking was rapped down diagonal, and we had a platform to work upon. A base for the adobe banco in the niche between the west wall and dining room passage was provided by many close-set 2 X 8s, which Tom assured me would give enough support. "Oh, yeah." Tom's assurances were usually quite wobbly, but this time firm.

Adobe work started in June. I couldn't get all the adobes I

needed from Sixto Montoya; so I scrounged around, trucking them in 50 at a time, sometimes six trips a day into town and out. The living room fireplace was built up in mesh with walls. The huge slab of hearth was established, with an opening left to an enormous pit, big enough for generations of ashes. Francisco, Ann and I followed Kitty's plans pretty closely, except that we tracked and adjusted the arch of the opening as Baca built it up, until it gained a certain beehive shape, wider than Kitty's. How Baca supported those curves of firebrick laid on edge, as they rose and leaned inward, how he cut the beveled edges of the key-brick of the arch, I don't remember. At the same time he raised the adobe masonry to embrace it, meshing firebrick and fire-clay mortar with the mud-work. The back wall rises vertical for a few courses, then it too converges inward to form the smoke shelf. This time we made sure to build in a damper to close the chimney.

Once the fire-hive was completed and enclosed in walls, a long angle-iron frame, welded according to plan, was laid in horizontal above the opening to serve as the mold for a plaster mantle. Why not wood? Don't remember why not; seems to me it would have been simpler to do, especially the two curved "supports," laboriously formed of lath and plaster, which are only there for looks. Francisco was a wizard at molding lath and plaster; he did it just fine, as he did the archway leading to the portal.

Major questions to be answered as the walls rose: who could make the five special windows for the dining and living rooms; how to raise the living room-to-portal doorway to incorporate a fan window; who could make a fan window?

In some places, I would again use Andersen casements, in various sizes and arrays. But the big double-hung windows baffled me. The summer before I'd measured the de Ghizes' windows that we wanted to copy, so the openings were known and planned for. But I couldn't find anyone to make them. It seemed to be partly a problem of cutting a molded edge: who had the equipment? Each half of the window had a single vertical divider, making two large lights in each, four in all. It seemed simple enough, to look at it, but no one in Santa Fe made windows. At last Tom ventured that his brother Carlos could make them. I was so relieved that I threw the job at Carlos, even though he wanted $80 apiece.

A tangle of problems arose with my inspiration to put a fan window over the French doors to the patio, but they were all solved well enough. The flash hit me well before the walls got up that high, so we could figure out and plan the change in good time. To raise the doorway, the bond beam on that wall had to be elevated higher, which incidentally gave a nice pitch to ceiling and roof from east wall to west wall. Likewise, the portal ceiling had to rise to an even steeper pitch—nicer yet. None of this was in the Bartell plans.

Bob, the manager at Big Jo, told me to try a man named Waddell for the fan. He was an elderly craftsman who lived in an obscure corner of town; I found him and he agreed to take on the job. What he accomplished was not exactly Georgian, but good enough: four spokes radiating from a hub, making five wedges, joined into a large semicircular whole, divided by a smaller semicircle. One side of the sash was rabbeted to seat the wedges of glass, which I had glazed in. It looked heavier than what I'd envisioned but the work was well done, well joined where he'd pieced together segments of the semicircle. In width, the sash matches the sash of the French doors. The vertical line of the door-sash almost lines up with the line of the smaller semicircle, so that to the eye they seem to run continuously. Nice, the way that happened to work out. The fan had no frame of its own; it was centered over the pair of doors and Francisco lathed and plastered smoothly right up to it.

Carlos Weatherford and the double-hung windows were another story. As time went on, he brought out elements as he finished them. I was disappointed to see the knotty common lumber he was using in the frame; I thought we'd agreed to use clear stock. The sash seemed crude, heavy, graceless, not the fine work I'd hoped for. His answer to the problems of molding: take a router to the inside edges. They came out not molded but rounded, even the corners, which were not fitted but rounded by his machine traveling all around the sash after it was assembled. He might as well have used a sander. To butt one molded piece perpendicular to another would have required a tool or a skill he didn't have. I had to settle for this—Santa Fe native carpentry, c. 1965. Ten years later, even five years, things would have been different.

The $80 included glazing and screens. The large size meant a lot of glass, which made the windows heavy, too heavy for the balances (raising and lowering mechanisms) then available in town, which Carlos got: plastic tubes encasing springs, which fitted into grooves in the edges of the sash and fastened into the frame. There was always trouble with them afterwards. Too loose or too tight, drafty, rattling in their stops, their tracks. (Windows have their nomenclature: frame, stool, sash, muntins, lights, rails, stiles. The check-rail of a double-hung window consists of the top rail of the inside sash and the bottom rail of the outside sash, which meet in the middle and close the window. The stops, "blind stop" and "parting stop," are narrow strips dadoed into the frame to form tracks for the sash to slide up and down. The stiles are the two vertical members of the sash.)

This work was beyond Carlos, but he charged and I paid. The screens were included—I thought. Before I gave him the final $50 I owed, I asked him where were the screens? He blanked. Screens? Screens were not part of the deal. I tightened up. He tightened up. No violence erupted, but Ann said later that we circled each other like cats. He gave way, or seemed to, and went around measuring for screens. Then he went away, and I never saw him again. No final payment, no screens. Tom said, "He won't come back. If you argued, he won't come back." Tom wasn't scolding me or taking sides, just saying what he knew to be true.

Francisco and Pipas kept gobbling up adobes, faster than I could truck them. Both were laying now, which raised the pressure on supplies. Finally, I got 1,200 from old Mike Barela at 8¢, and 1,200 more from ancient Luis Lucero at 10¢, and paid Armijo to deliver them. That held us until Montoya got his production going again later in July. He said he couldn't find guys to make them. He probably paid a dollar an hour, so no wonder. "I gonna make you a good deal."

With two men laying, I needed more help to make the mud and wheel the adobes. Francisco brought his son Philip—"Felipe"—"Phil."

At 20, Philip was an amiable giant who walked with a relaxed, slope-shouldered slightly hunched gait. He revered his dad, but he was a fine fellow himself. He had a smiling mouth, narrow eyes, white teeth, dimples and mole or birthmark on his cheek. He had a Buddha-like smile. He seemed always gently amused; had the easy motions and assured power of a man who could break you in two with one swipe. He told how he got in a fight over his wife one night and how he beat up the other man. "I kept kicking him," he said in surprise at himself. His head and face were so handsome he might have stepped out of some heroic comic strip. He was suffering from "family trouble," he told Ann. He and his wife had lots of trouble, were separated part of the time he worked for me, and eventually were divorced.

He stood about six feet and was so strongly built that tools looked puny in his hands. And yet there was the dreaming, smiling quality about him. He daydreamed while he worked. Sometimes his ear was plugged into a little radio that only he could hear. He liked painting windows because it was conducive to dreaming. He seemed to have an extra eerie sense and swore to me there was a place where a car would roll uphill.

I remember his placid, easy motions pitching shovels of dirt through the propped-up screen, hoeing the mortar-box, which was half full of mud, half of dirt soaking. He worked bareheaded. He admired his father, yet it didn't make him restless like his father. Ten years later, after Vietnam and the Marines, he looked hunched, his eyes drooped at the corners; he seemed somehow smaller, shrunken, much older. He said he still crouched at any sudden noise, as if he were still amid danger.

But back in July of '65, he went about the world gently, his face secretly beaming, his mind wandering far away from shoveling, mixing, wheeling.

Nearby, I worked at the aspen poles to ready them for latias, which this time I made sure were no thicker than two inches. All those in the front hall and around the south hall are just right.

It was exciting to find the poles. I went once with Sam Pratt to get a bunch, once with Bob Kurth. Kurth wore an old raggedy hat, and he was so strong he lugged armloads of poles uphill from the down-side of the road. The best were still standing, dead but not rotten. Thin columns—shuck off the loose bark and reveal the beautiful color on the smooth round bone of tree: ash-gray, smoke blue-gray, bluer than smoke, wintry blue, snow-shadow blue. Dapple brown, blue-and-brown piebald, brown as sandalwood or juniper heart. It became a brown and white creek flowing above the vigas in currents; eyes of knots appeared in the stream, in the flow of round smooth sticks faintly undulating as muscular limbs. Trimming the poles exposed beige or pink-beige or off-white tones. I made slashes like brushstrokes against the dark gray.

I worked for days on end in the July heat and sweat, with tiny gnats and flies swarming at my nose, eyes and ears. I stood the aspen against my shoulder, whittled and husked them with a hatchet and a cheap tin-hafted knife. It was fruitful labor. I keep looking at them up in the ceilings. The streams. Let them sing for themselves.

We wanted split red cedar in the ceiling of our front entry porch, or "vestibule," so Ann and I went up the Rowe Mesa to find it. It was paradise to me, that other country, that other world at the top of the mesa. There I found wind, solitude, tall trees, ponderosas and giant piñons. Lonely, two-rut roads wandered for miles through a wild of hills and valleys, ruts so ragged and rough they seemed meant for wagons or jeeps. There were thickets of dwarf oak, russet and ochre in the fall. At the rim came an enormous valley-sweeping view, where we could follow the Santa Fe trains and tiny flashing cars, and look across Rowe and Pecos to the mountain ranges. After everything else decays, after the details of living in Santa Fe have palled, it is the magic of that country and that light that always make me wish I were back there when I'm away.

On Rowe Mesa I found pitchy pine-stumps and knots, treasuries of kindling-wood. There I found red cedar, straight-grained and straight-splitting. Green cedar is good for posts, dead and dry is better for latias. When first cleft, the inner heart shows beet-red, beet-purple, and gives off a luscious aroma. In time, it fades to red-brown, nearer in value to the adjacent streaks of white that fade toward yellow and brown.

The Mesa is the home of magical sandstone, stones of paradise, stones like gold in the woods. It is fine-grained stone, easily worked and melted by weather, but wondrous while it endures. Going through in May, hunting for cedar, I crossed a ridge to the left of the road and found myself in a wide, pleasant valley of long grass and forested heights. The woods opened up in a hidden vale. I noticed some loose stones lying around under the trees. So they are discovered, odd glowing stones here and there on the floor of the forest, a clutter here, an outcropping there. I lingered over these. Gold, salmon, pale brown they were, fine-sandy and sparkling tinily, edges and points worn round. Then I passed on. But I never lost them. They lay on a certain slope, under tall trees, and light and shade fell on them in a certain way.

Though I went on and got what I was after, drove down off the Mesa and twenty miles home, I never lost them. They glowed in my mind. A long time later, I went back to get them, to find them again. And I found my way. They still lay there on that slope of valley, shaded by the tall, wide-spaced trees. I took those and many others, not only the salmons, golden browns, yellows and reds, but grays and olives spotted and marked with black calligraphic strokes, and others starred with tiny blue lichens. Most of those wild softly-glowing stones from the Mesa are now steppingstones that I planted in the patio terrace for grass to grow around.

The patio was a dump, a wasteland, when we began to build the portal in August. We started with spot footings to carry three columns, and two small adobe bearing walls on either side which Baca wired to the existing structure with lath. Heavy rods were stuck in the wet footings as dowels for the heavy wood columns, which Tom drilled to fit down over the dowels. At the same time, the pillars had to be lifted up to let the future brick floor flow in under them. I went to a welder for "column jacks," short lengths of pipe with flat steel flanges like feet welded to one end. The pipe would fit down over the dowel, the flanges would hold up the column.

To top each column, a capital, which Tom sawed out of an 8 X 10 timber after Kitty's angular tree-like design. Bearing on capitals and columns, a 35-foot, patio-wide architrave of 8 X 10 rough-sawn timbers, which were coupled over each column in a scarf joint that looked like a "Z" on its side. This jointed beam would carry the portal joists, the weathered 4 X 6s I'd bought the previous fall.

In the garage there were no such refinements. When the adobe walls rose to door height, we laid a large timber lintel across the opening. At roof height we put 2 X 12 rough-sawn rafters 16 inches on center, with rough decking to cover them. We chose to leave adobe work unplastered inside, the only place in the house where adobes were exposed. To hold the 2 X 12s on the finished hard-plaster wall that was already there, Tom said, rather than dig them in, we could set metal brackets for them in the wall with a "ram set," a gun that shot nails. I found brackets ready-made at Big Jo and rented the ram set. Each cartridge for it was tipped with a nail and was color-coded yellow, blue, et cetera, according to power of charge. You loaded in a cartridge, put it up to the mark and *blam*—it blew the nail right in.

Francisco and Tranquilino graded and finished the slab floor and apron, leaving a row of bolts sticking up in back to anchor a 2 X 4 partition for a store room, to be sided with vertical rough-sawn planks.

Also vertical and rough-planked was the pair of doors I designed with boards braced across the back in a Roman X, crossed at top and bottom. The only refinements: lumber was trued on the joiner, and Tom put everything together with blue screws and washers. It must have abraded even his tough hand, all that screw-driving. Hung on huge black strap-hinges, those doors were bound to slacken, wood to shrivel, screws to loosen, as they knocked and knocked against their outside moorings. Windstorms would yank and break them. I can still hear them chafing, bumping, rattling in the least breeze. Even so, I liked the way they looked.

The portal joists that rested one end on the architrave were let into the outside kitchen wall and into the chimneybreast by chopping a hole for each. How I used to look forward to seeing how they and their weathered decking would look, up in place. I was not disappointed. I gloried in them and in the brick floor beneath and in the cedar shake-paneled doors of the closet.

And wonderful to see the cedar shakes woven in Vs over

the living room 6 X 10s. And when my specially trimmed latias were cut to length and laid up over the round vigas I had whittled and stained, my cup filled up. Francisco helped Tom nail latias in place—this time we finished them before roofing. At this angle of the long corridor, by the front door, we turned the roofing square, with dark 6 X 8s to carry the vigas. Beyond, the vigas resumed at right angles. No wheeling of vigas possible here, the front porch cut into the space.

Opposite the angle was the passageway to the guest apartment; the lintel across that opening carried one of the 12-foot natural vigas for that section. I left those logs as they were, peeled but unstained, so their own straw-yellows, browns and moth-grays would cross and enhance the darker, bluer tones of weathered pine decking above. I wish I had just that one room and ceiling now, to lie under and look up at.

Split red cedar went up over vigas I left untouched, because of their purple streaks of bark, rich as a Persian rug. Once the ceilings were all up, shakes and latias had to be paneled over with plywood for the roofers. Walls were vulnerable now; rains came and cut through between roof-deck and adobe, plowing out the mud. Early in September, Farwell's troupe roofed us over and rescued us from further devastation.

Around that time, I called for the ready-mix truck to pour the garage floor, the front entry, and the terrace step footings. The adobe parapets were built up to height, with metal flashing to protect the roofing. Now began chicken-wiring and plastering, which went on into October. With the front porch protected from weather, I had Francisco finish the walls and the banco with mud. We would leave the floor rough, as bedding for a future pebble mosaic (never accomplished).

The mighty front door, which I had spent hours dreaming up and designing, had been given to Werner Hildebrandt to build. Now it was ready. He brought it in two huge slabs of laminated walnut, to be hung as halves of a Dutch door. Three plies of ¾-inch walnut I wanted, each layer running in a different direction to help the door from warping. The inmost layer, to face the interior hall, was to be of boards vertical and side by side; the outer layer, facing the outside world, was to be set in twin patterns of squares receding inward. Each square or quadrangle of wood was to be a different width from the next. Werner carried out the design beautifully. It looks like inlay, like the intarsia perspectives you see in Florence and Siena. Inside, he matched the grain of vertical pieces so that it appears continuous, top to bottom. The oiled darkness of the walnut, streaked with yellow, gives it a special richness and majesty. Outside, darks and lights of wood form themselves into mitered light-struck halls receding toward far walls that look like stone. So beautiful!

It was Werner's triumph and mine. And Tom Weatherford's. He devised the overlapped casing that leads into the door, the setting for it. To extend its concentric design, he matched a series of clear pine boards of varying widths all around the flared opening and rabbeted their adjoining edges to fit together in what he called "shiplap." I stained it walnut and I oiled the door with linseed oil.

Never mind the trouble of finding hardware to hang the halves and lock them; never mind that Werner later offered my design in his catalog without asking me or giving me credit; never mind that winter sun strikes the lower half, fading it and separating the joints. Still it is splendid; it rejoices me to think of it.

Before Tom put up his casing, Francisco laid the doorstep of sandstones that I got from Casey Willard, and which we also used to build the terrace steps. Casey Willard was the local salvage baron. He was a quick, busy man in glasses, dressed in khaki shirt and pants and a Penney's khaki snap-brim hat. He spoke with a rapid Texas speech, nervous, no drawl. "That's the same identical brick as the others," he told me when I wanted a comparison. "Very same identical brick."

He was a packrat. In his yard on Agua Fría Road were mountains of salvaged brick, stone, flagstone, concrete and junk of all sorts. He sold the sandstone we used, cut about three inches thick, ash-gray, buff, golden in hues, for $7.50 a yard, and he'd gotten it for next to nothing. He grabbed for a pittance material and rubble from demolitions, and sold it dear. Ed Gerhauser, now working for Willard as a partner, told Ann and me that Willard salvaged brick from the demolition of the former Quick Car-Wash on Alameda and sold it back to the builders of the Forge and Inn of the Governors now standing on the same site. Willard sold used brick at different prices, none for less than 5¢ apiece. (New common bricks sell for 5½¢, 6¢ apiece.) Some he sold for 7¢, some for 15¢. Extra-hard black paving bricks—"penitentiary brick"— went for 15¢.

At intervals over the summer, I went to Willard's and picked and hauled brick to put in our portal floor. It was tedious. I picked through his stacks, discarding what I didn't want. Stacked them in piles of 25; toted those piles to the truck in armloads of six or seven; dumped them on, then laid them on edge so they wouldn't break over bumps. On and on, one of those ant-projects. They clinked like pottery.

As plastering went on and weather got colder, the exterior wood needed painting. A story in the *New Mexican* caught me: a family named Moreno was ruined by fire, all their possessions lost; the man, Gabriel, was a house-painter and was looking for work. I called him and offered him the job. He was eager and competent; he painted for a week, before moving on to more jobs his bad-luck story had won for him. For the rest, the interior kalsomine and woodwork, I hired Gus Bokum and his crew again.

Inside plastering and flooring were still to be done, before the painters could move. The fireplace, the corner banco and niche, the bookcase and the archway to the patio were the special problems of this part of the house. For the archway, Francisco worked out a lath framework that started on a wider radius at the inside because of the flare of the walls, and that narrowed as it converged on the fan window. I recall him nailing a board across the opening to give a fixed point for a wire that he swung radially to determine the arc. He extended his horizontal guide to the fan by fastening another board perpendicular to the first, making a T.

He and Tom installed furring to cover the gap between ceiling and bookcase and the gap between ceiling and banco niche. The small lamp cubbyhole in the adobe niche had been built in; the niche was linteled over with timbers and enclosed by the furring. To form the mantle and its brackets, Francisco bent metal lath around the built-in angle-iron frame and formed lath in two quarter-circles below it, and plastered it all over. He smoothed a hard cement finish over the raised hearth, which I ultimately painted blue-gray. The rest of the chimneybreast was finished with fine mud and whitewash like the other walls.

Except for the bookcase wall, the partitions of the guest apartment are stud walls, which had to be lathed and plastered. Same with the closets, one for coats, one for water heater, which are back to back and share the same viga, but are two feet different in elevation. Guest quarters are on the same elevation as the living room, but are approached from the entry hall, thus the four brick steps down.

The steps lead to a cabinet, bar, sink and tile counter built in below the small six-light casement. We flared not only the window, but the entire cabinet enclosure: one reveal inside another, recalling the design of the front door, also the stepped dining room embrasure. At this distance, I can draw their shapes in the air with my finger. With hindsight, this cabinet, with its panels by Hildebrandt, seems a waste of precious space. The bathroom is too small; more room should have been given it. The bedroom is snug if not tight. No closet or cupboard was planned for it; so years afterward, Tom and I built in the wooden corner wardrobes, perhaps pinching an already small space. (I start to rebuild and move things around in my mind, when reality pulls me up short—it's not mine to do anymore.)

I decided we could handle the flooring ourselves this time, rather than give it to the professionals. Before flooring, we broke through the temporary bulkheads in the living-dining doorway and at the hall angle. They were hard exterior plaster over wire netting over stud frames, a fierce wrecking job. But now the floors could join and the two halves of house could flow together. Floors met without a hitch at living-dining, right on the money. At the southeast corner things got out of line, so there's a slight jog in the wall there and maybe a slight blip in the floor, too, though I can't remember.

Once the subfloor was shoveled and scraped and swept, a dry sheet of red building paper was laid down over it, and the joist-lines were marked on the paper so that the oak could be nailed through subfloor into joists. Special spiral flooring nails were driven toenail through the top and back of the tongue of the oak board, with a device like a giant stapler that set each nail at the proper angle and fit in against the tongue. Pound it once with a big mallet and it rammed in the nail, all the way.

It was Kimball who advised me how to lay the flooring. Don't use up short lengths in ending a row, he said; cut the ending piece out of a long board and use the remainder to start the next row. Each board came with a tongue across one end and a groove across the other. Complete short boards would be needed somewhere, so don't waste tongue or groove by driving it under the baseboard. Cut it out of a long piece, stuff the blind end under the base, and use the tongue to match the groove of the last board. Then turn over the remainder, stuff the blind, or cut end under opposite baseboard, and use the grooved end to start your next row of flooring.

Hard as stone, the heavy white oak crashed and clattered in the rooms, which echoed like a warehouse because they were bare of fabrics. Even the small scraps banged when dropped. Sharp claps of board on board, thunderous din of bundles. Final rows were forced in under the baseboards with a block and jimmy, or were ripped to fit and nailed down through the face.

Sand, fill, seal, finish. Each stage of flooring was difficult. The big rotary sander sounded like an airliner taking off in the room. Lean back on it, restrain it, brake it or it would run away with you. It took know-how. Any pause or hesitation and the sander dug a groove. Jimmy Jenkins never made a dent; we made quite a few. I contributed my share. Once when I missed a nail, Tom said in a funny little voice, "I'm gonna fire you!"

There was a knack to keeping the sander progressing evenly. We went over the wood once with a coarse belt, once with a fine belt. From time to time, we emptied sour spongy dust from the bag; from time to time a belt would pop like a gunshot and the grits would fly. We used a smaller disc sander for edges and corners.

After sanding, we used a paste wood-filler, mixed with thinner, and brushed it thick across the grain and cracks. We scattered in oak-dust, then scrubbed away the excess by furious cross-grain wiping with burlap. Then we wiped syrupy sealer across the grain. Then we steel-wooled away the excess. (Two coats for this, twice over.) Then we waxed and buffed. It was one hell of a job! Tom, Philip and I did it. Compare our work with Jenkins', amateur versus professional work.

By the same method as before, same bricks in the same basket pattern, Francisco laid down the dry brick-on-sand floor of the hall and cemented the brick steps. Not long afterward, it grew apparent that the bricks near the entry were sinking. The earth fill at that point was so deep that it continued to settle, and with it the sand and bricks. Bricks are so tight together that to unlock them, you have to break one out. About 10 years after the building, I got Francisco to pull them all up, repack the sand and reset the brick. But until the dirt below ceases to settle, the brick will go on sinking.

He got one set of steps wrong, and I had the dreadful task of telling him, tear it out and do it over again. He took it: "Well—okay," and undid his work. The peril of setting him straight! He was mostly right; he knew so much; he was proud of his work.

Well he might be. The portal floor is another of his memorials. And Ann's. It was her design and she supervised his work. Her pattern was inspired by a floor in the old Gilbertson house on Alto Street. And she had been taken with the appearance of a stack of used bricks she saw by the road one day, the effect of dark checkered with white, of bricks powdered white by plaster or paint. Francisco did the work; his the labor and the suffering. He had to cut nearly every brick to make the curves of the design. It was November and cold, and his knees were miserable. We put a row of black pavers along the outer edge, on a concrete and block footing. Inside that retaining wall, the bricks were laid on edge, in sand. A great rugged mosaic was the result—intersecting arcs and fans of variegated reds, blacks, blues, whites. This lively, flowering brickwork is the glory of the portal, which is my favorite area of the house.

Finally, a progression of finish: painting, electrical lamps and outlets, plumbing fixtures, tile in the bath, door hardware, louvered shutters for the living room windows.

Closest to my heart was the final cabinetwork. For the dining room entry, Werner delivered a pair of pine doors with arched panels, which we designed. Tom Weatherford built the bathroom cabinet and the drawers below the window. I proposed and he figured out and built the doors to the portal closet. He took weathered two-inch rough-sawn lumber, rabbeted the edges for panels, pegged together the stiles and rails with dowels, and faced the panels with cedar shakes. (Stiles, or verticals, and rails—horizontals—compose the frame for paneling.) These doors were my special pride and joy. In the same vein of gray rough lumber and shakes, he built in the cupboard below the fireplace.

But his finest work was the bookcase and its fluted 1 X 6 trim. He beveled panels and set them in the cabinet doors, and joined the mitred corners with dowels. He dadoed the shelves into the sides of the bookcase. And, to answer my craving for 18th century woodwork, he channeled out the parallel grooves of the trim with a router. Long ago, when we first began to build, Tom had told me this was his favorite kind of work. "It's clean work," he said.

All the lumber is good, clear pine, which Philip thoroughly sandpapered. For paint, I used a velvety warm gray, a flat soft McMurtry that was a pleasure to brush on. When finished, it smelled good and felt good. Nice fluting for fingers to run along.

Except for subcontractors, the work petered out as we entered December; so one cold gray day I took Philip Baca out on an expedition to get firewood. At his direction, we turned in before Cañoncito, at the western end of Rowe Mesa, which is also called "Glorieta Mesa" or simply "the Mesa." We entered a rolling land of piñon and juniper and occasional ponderosa. Phil took me to visit the rancho, where he said his dad had been raised, and which his uncle now owned. Phil loved the place and showed it proudly: "Well, how do you like my little ranch?"

There were chickens, turkeys, guineas, ducks, pigs, goats, horses, a cow, a bull—one or a few of each. There were rough pens and sheds, but the chickens ran loose. Philip showed me a small chinked log-house where he said his father had been born and raised. It was a ruin with a relic of a stove inside. "Log Cabin" was no myth in the New Mexico of 1920, out in the wild of the Mesa.

There was an earth-banked reservoir to catch water runoff. The main house had been built of ugly block by Felipe's uncle, and we ate our lunches there with the trampy young man who was the hired caretaker. He was a queer, unshaven type, a dreg, who stayed full-time alone there. It was he who had painted and carved the hideous face on the tree-stump as a hex to ward off evil spirits.

After lunch, we went off to hunt for more wood. We found a big fallen tree and decided to cut it short enough to get it in the truck.

"You chop funny," he told me. "My dad showed how you have to make a chip." He hacked a square notch, not a V as I'd been doing. The square cut made a big chip, so the axe wouldn't get trapped in a narrow slit. There was a way to do everything, Phil knew from his dad. As I'd already discovered, Phil knew the kindling richness of pitchy pine, resin-saturated relics, which he called ocote. He would fix a shovel as his dad showed him: burn out the broken handle, set in a new one and soak it in water to make the wood swell. The trouble with this is, the wood always shrinks afterward, smaller than before, and the handle falls out. Never soak handles!

We made sure to spike our load with plenty of ocote. On our way home, Felipe told me how his dad and another were out firewooding on the Mesa. They had their load gathered and built, when a warden came up to them. He was suspicious and forced them to unload all their wood to prove they did not have a deer hidden underneath. It was the end of the day. They unloaded the wood. There was no deer. Then Francisco picked up the warden's rifle and at gunpoint ordered the warden to reload the wood. The warden reloaded the wood, and that was that.

Phil said his father told him and his brothers always to eat all they wanted, all they could hold; there might come times when there wouldn't be enough food, but while there was enough they should eat all they wanted. "Chicharrones!" Phil said in rolling, sizzling Spanish. "You should eat chicharrones!" Pork cracklings or chitterlings, which he relished.

By mid-December, I'd paid off all the men, and we went into hibernation for the winter.

One big job remaining was to stucco the house. Francisco and Eloy Padilla accomplished that in April 1966, laying a seamless film of "Adobe 85" over everything. Then, oh then, what a wonder we had! Walking out among the green piñons, I would look back at the house and see its rounded parapets and walls rising all warm and earthy red-brown above the green. The dreary gray was all swallowed up—"into something rich and strange." In a certain light, the color was ochre; in another, salmon; in another, cinnamon— very like that changing Rowe Mesa stone.

Another big task awaited within. Full of rubble and scraps of stucco and cement, the patio looked like a prison yard. In May, after the stuccoing and after a trip to Juarez with Ann in the pickup truck to buy pots and planters and outside furniture, I cleaned out the patio, shovel by shovel, wheelbarrow by wheelbarrow. And by the same method, I brought in topsoil, graded and landscaped; I channeled in a drain to run overflow out through the piñon-tree well and out under the house. Then we planted rose and forsythia and grass, and put a little slip of a Russian olive on the terrace amid the wild Mesa stones.

The long struggle was over. It surprised me then, it surprises me still, after being engulfed for so long in mess and jumble, how presentably the whole thing came together. It turned into a real building. That is partly the blessing of finish-work, which smoothes over all the cracks and botches and ragged edges. The first years of life there, I had mixed feelings, not all happy. I knew where the mistakes were, I could see through the finish.

Now, the farther I get away from my house, the more beautiful and wonderful it seems. Maybe this writing will help me say goodbye to it. I keep missing it.

June 1966 to July 28, 1986

Postscripts

In the fall of 1966, I built myself a tiny adobe house (studio) out near the corral. A year later, we hired a contractor and added Ann's studio onto the master bedroom. Its north wall is adobe; its east and south walls are wood frame and rough-sawn siding. Some time after that, Mike Lujan came into his own by building us a round pueblo horno, a bake oven, out in back of the kitchen.

In August of 1975, we left our house, intending to be gone at most two years; but as it turned out, it was forever. We never lived there again. In July 1979, during a period of separation, we sold it. Two years later Ann and I were divorced.

Big Jo is gone, superseded by the El Dorado Hotel; Santa Fe Builders is now the Sanbusco Center, and some people wonder who San Busco is, which saint.

Perhaps Francisco still lives, perhaps old Tranquilino still lives at 85. Tom Weatherford has died; Eloy Padilla and Mike Lujan have died. Last I heard, Willie Sena was State Plumbing Inspector.

HOUSE BORN OF MUD

See how the weeds and lizards board
The bleached gunnels of mortar-box beached
On a crackled hill, that one time plowed
Seas of dirt, lugging and cradling
Cargos of *mezcla* batched by a mud-cook.

Hoe for a spoon, he chopped a batter
Of earth and water, sand and straw,
Drove it and hauled it until it slapped
A luscious mush that wheelbarrows ferried
To scaffolds of bakers building a cake
As big as a house.

                    Eager to home
Its creators came for the salvage or solace
Of newborn rooms. The blank of the future
Was building a shape before their eyes.
The children would make the place their own
By play and fear and window theaters,
Their refuge from school, palace of Christmas,
Wedding chapel.

But time would come—
What jolt to them then to know
How short their time would be—which in
Its slow exploding and dispersing
(No time for a wedding) they'd accept
As right, when they would leave their house,
Their harbor, and go their separate ways.

Born of mud and desire for beauty
The house would stand a home or a stage
For others, maybe for generations.
All of it hatched from that wreck of a cradle
Beached on the side of a burnt-out hill.

—W. N. Gates

www.ingramcontent.com/pod-product-compliance
Lightning Source LLC
Chambersburg PA
CBHW032101080426
42733CB00006B/371